Biology of Freedom

Biology of Freedom

Neural Plasticity, Experience, and the Unconscious

François Ansermet
Pierre Magistretti

Translated by
Susan Fairfield

KARNAC

First published in 2007 by
Karnac Books Ltd
118 Finchley Road
London NW3 5HT

Production Editor: Mira S. Park

Printed in the United States by Other Press

British Library Cataloguing in Publication Data
A C.I.P. for this book is available from the British Library

ISBN: 978-1-85575-606-9

Printed in Great Britain
www.karnacbooks.com

Consider that for a while each single living creature is said to live and to be the same; for example, a man is said to be the same from boyhood to old age; he has, however, by no means the same things in himself, yet he is called the same: he continually becomes new, though he loses parts of himself, hair and flesh and bones and blood and all the body. Indeed, not only body, even his soul, manners, opinions, desires, pleasures, pains, fears, none of these remains the same . . . .

Plato, *Symposium,* 207d-208a

Contents

Acknowledgments

We want to thank Bertrand Cramer and Bernard Golse for challenging us to explore the possibilities for dialogue between the neurosciences and psychoanalysis, and Leslie Ponce, a friendly and analytical witness. We are also grateful to Catherine Cornaz for her patience and diligence in transcribing this manuscript and to Sylvain Lengacher for his availability and skill in the preparation of the diagrams.

Preface

This book came about through a meeting: a meeting of two domains, psychoanalysis and the neurosciences. And a meeting of two people as well: a neurobiologist who had a personal psychoanalysis and a psychoanalyst open to what other fields can teach psychoanalysis. And, finally, a meeting based on a mutual observation, namely, that experience leaves a trace.

This observation has found experimental confirmation through recent progress in neurobiology, which demonstrates a plasticity of the neuronal network permitting the inscription of experience (Kandel 2001a). This plasticity, which nowadays is believed to underlie the mechanisms of memory and learning (Malenka 2003), is fundamental for neurobiology. It has allowed us to leave behind a rigid view of the nervous system. We now take for granted that the subtlest elements of the process of transferring information between neurons, that is, the synapses, are permanently altered in accordance with lived experience. The mechanisms of plasticity operate throughout a subject's life and significantly determine his future.

Although the experimental results demonstrating the existence of this plasticity are recent, the hypothesis itself is an old one. Santiago Ramon y Cajal had already formulated it over a century ago (1909–1911, 26): "Thus neuronal connections are not definitive and immutable, since, as it were, provisional associations are created that are destined to remain or to be destroyed according to indeterminate circumstances, a fact that, incidentally, demonstrates the great initial mobility of the growth of the neurons." Freud himself (1895) grasped its role in the mechanisms of learning and memory, and the hypothesis was revisited several times, especially by Donald Hebb (1949). In other words, the conceptual soil was ready to receive the experimental data. The Nobel Prize in Medicine for 2000, awarded to Eric Kandel, was a major recognition of the importance of the mechanisms of plasticity in modern neurobiology (see Kandel 2001a).

The idea that experience leaves a trace is also central for psychoanalysis via the concept of the mnemic trace left by perception and its different levels of inscription, whether conscious or unconscious. The originality of Freud's hypothesis was its assumption that there was not one sole inscription of experience, but that experience was transcribed in different systems, finally ending in the constitution of an unconscious mental life. At that time, all Freud could do was intuit what biology was unable to confirm: "The deficiencies in our description would probably vanish if we were already in a position to replace the psychological terms by physiological or chemical ones. . . . Biology is truly a land of unlimited possibilities. We may expect it to give us the most surprising information and we cannot guess what answers it will return to the questions we have put to it" (1920, 60).

Have we today, at the beginning of the twenty-first century, reached a level of knowledge of biology that would allow us to objectify the nature of a trace produced by experience, sketching a bridge between the psychic trace and the synaptic trace established in the neuronal network?

In neurobiological terms the trace is dynamic. It is subject to modifications. The mechanisms of its inscription give the neuronal network great plasticity in the original meaning of the term. This is how, on the basis of experience, an internal reality is constituted. This reality may, of course, be conscious, underlying the memories we can recall to awareness, but it may also involve inscriptions, from the domain of the unconscious, that we cannot evoke. One of the major topics of this book is precisely the exploration of the mechanisms permitting the establishment of this unconscious reality, with emphasis on how it influences the way a subject evolves. In a heuristic approach, one might even say that we have tried to outline a model in broad strokes, one that is subject to debate but that will serve to explain the biology of the unconscious.

Traces are inscribed and linked; they vanish and change throughout life by means of the mechanisms of neuronal plasticity. These traces inscribed in the synaptic network will thus determine the subject's relation to the external world. Hence they affect what becomes of him. This is an important point to stress, since it implies that the subject is continually changing, setting out each day from a *tabula rasa* on which new traces will be inscribed. This raises the question of maintaining the subject's identity throughout his history. After all, the mechanisms of plasticity, as described by neurobiology, involve the constitution of a lasting, if not permanent, trace. Plasticity is not synonymous with flexibility[1] or permanent adaptability that would leave the subject without a certain determinism and a certain fate that is his own.

Plasticity plays a role in the emergence of the subject's individuality. Each of our experiences is unique and has a unique impact. To be sure, plasticity itself entails a form of determinism, but at the same time as it produces this form of determinism of the subject it frees him from genetic determinism. For, if we bring experience into play as a determining factor in the subject's process of becoming, we distance ourselves from an exclusive genetic determinism that fixes the subject's development from the outset. Thus plasticity is no more and no less than the mechanism through which each subject is singular and each brain is unique and free. Hence the title of this book: *Biology of Freedom.*

We could also have entitled it *Sculptures of the Unconscious* in reference to a sculpture, made in 1930 by

1. See the critique by Malabou (2004).

Alberto Giacometti and called *The Hour of the Traces*, that strikingly illustrates what we have just said. In 1930 Giacometti stated that he created this kind of sculpture without asking himself what it might mean; but, he explained, "now that the object has been constructed, I tend to find in it the transformations and displacements of images, impressions, facts that deeply moved me (often without my knowledge), forms that I feel are very close to me though I am often unable to identify them, which makes them still more troubling to me."[2]

Giacometti seems to have created this sculpture in an almost automatic fashion from elements of his unconscious. Its installation can indeed be seen as a metaphor of internal unconscious reality constituted trace by trace, in an almost makeshift manner, where unexpected elements are linked according to the subject's experiences and the responses stemming from the uniqueness of his own psychic life.

The second major argument we shall be setting forth in this book is that the constitution of this unconscious internal reality based on the mechanisms of plasticity is not exclusively a mental phenomenon but involves the body as well. For we shall be discussing the fact that the traces left by experience are associated with somatic states. Our argument is that the perceptions

2. Cited in Sylvester 2001, 74. The description of the sculpture at the Tate Modern in London notes that this fragile construction suggests the mysteries of the unconscious, combining space and time, eroticism and death.

leaving a trace in the synaptic network are associated with a somatic state.

This claim is based on a whole set of recent data in the neurobiological literature, synthesized by Antonio Damasio (1994) in the theory of somatic markers; it also leads back to the first hypotheses of the origin of the emotions proposed by William James (1890) at the end of the nineteenth century. According to this theory, perception is associated with a somatic state; the recollection of the somatic state associated with a perception brings about the emotion. Perception alone, in this view, is neutral with regard to emotion. The reading or recollection by particular neuronal systems of the somatic state associated with the perception, or with the traces it left in the synaptic network, is a determinative factor in subjective emotional experience. In light of the theory of somatic markers, we shall revisit the concept of the drive, defined by Freud (1915a) as a frontier concept between the somatic and the psychic. This will take us beyond the perception–emotion relation to the relation of unconscious internal reality to the somatic states associated with the elements constituting it.

Having set forth the biological fact of plasticity, the convergence it implies between psychic trace and synaptic trace at the interface between the subject and the organism, and having explained its role in the emergence of individuality, this book will offer hypotheses for a model of the unconscious that integrates the recent findings of neurobiology with the foundational principles of psychoanalysis.

1

The Polar Bear and the Whale: What Plasticity Entails

At the end of his life Freud made the following observation: "We know two kinds of things about what we call our psyche (or mental life): firstly, its bodily organ and scene of action, the brain (or nervous system), and, on the other hand, our acts of consciousness. . . . Everything that lies between is unknown to us" (1938, 144).

Here we have the two terms of a debate involving, on the one side, neurobiological reality, and, on the other, the productions of mental life. These two fields, it must be acknowledged, have incommensurate features. A psychoanalytic colleague ironically compared our attempt to relate neuroscience to psychoanalysis with the unlikely coupling of the polar bear and the whale. And indeed, making any bridge between them can seem like an impossible undertaking, at any rate a dangerous one giving rise to confusions and aberrations that can only lead to the loss of the logics specific to the approach of each domain. The study of the brain and that of mental facts lead to radically different questions, implying fields of exploration and methods with nothing in common. If we consider individually the neurosciences on the one hand and psychoanalysis on the other, we see the extent to which these two domains are incommensurate and have everything to lose by being united in a vague syncretism that would make them forget their foundations.

A discovery on the one side is not necessarily one for the other side, and we are very far from knowing the relations of concatenation and causality between organic processes and mental life,[1] but the two are nonetheless intertwined, linked, in a global phenomenon. One day an account will have to be given of this enigmatic linkage, which Sganarelle notes so well in his own fashion in Molière's *Don Juan* (1665, 62): "My argument is that there's something wonderful in Man, which none of your clever scientists can explain. I don't care what you say. Isn't it wonderful that I am standing here, and that I have something in my head which makes me think a hundred different thoughts at once, and makes my body do whatever it likes?"

Until recently the same scenario was repeated endlessly between the neurosciences and psychoanalysis: one of the two partners in this impossible couple ultimately denied the existence of the other, excluding it for sev-

1. As Freud (1891) noted, this time at the very beginning of his work, the chain of physiological processes in the nervous system is probably not causally related to psychic processes.

eral decades. This occurred on both sides.[2] Over time, with rare exceptions, the result was either fixation on certainties and *a prioris* or speculative and confused debates. Drawing a caricature, we might find, on the one side, neuroscientists sure of themselves, most often reductionists, in quest of a biological etiology for mental illnesses, seeking a molecule that offered salvation. On the other side, we might find psychoanalysts most often rejecting the neuroscientists so as to defend their own concepts, to the point of getting similarly caught in the traps of reductionism, in any case accepting the split at the risk of becoming obscurantists.

Everything seemed to be definitively frozen between the neurosciences and psychoanalysis, with pendulum effects that, over time, successively favored one or the other. Breaking with this picture, the phenomenon of neuronal plasticity—a surprising fact emerging from recent findings in experimental biology—has completely overturned the terms of the opposition, bringing this pair into play in a new manner.

What the phenomenon of plasticity demonstrates is that experience leaves a trace on the neuronal network, modifying the efficacy of the transfer of information at the level of the subtlest elements of the system (R. G. M. Morris et al. 2003; Kandel 2001b). This means that, beyond what is innate, beyond everything that is given at the outset, what is acquired with experience leaves a trace

2. This was especially true on the side of the neurosciences, as is lucidly pointed out in Echtegoyen and Miller (1996).

that modifies what went before. The connections among neurons are permanently modified by experience (Blake, Byll, and Merzenich 2002), and the changes are both structural and functional. The brain must thus be thought of as a highly dynamic organ in permanent relation with the environment as well as with the psychic facts of the subject or his acts.

Plasticity introduces a new view of the brain. The brain can no longer be seen as a fixed organ, determined and determining once and for all. It can no longer be considered a definite, fixed organization of neuronal networks whose connections are definitively established at the end of the period of early development, setting up a kind of rigidity in the treatment of information. Plasticity shows that the neuronal network remains open to change, to contingency, that it can be modified by events and the potentialities of experience, which can always alter what has come before.

We shall return to the issue of what an experience might be said to be. For the moment let us stay with the fact that plasticity significantly changes our understanding of cerebral functioning and its relations with the environment and with psychic life.

Plasticity entails the obvious fact that, through the sum of lived experiences, each individual is seen to be unique and unpredictable beyond the determinations of his genetic background. The universal laws defined by neurobiology thus inevitably end in the production of the unique. The question of the subject as an exception to the universal now becomes as central for the neurosciences as it is for psychoanalysis, leading to an unexpected meeting point between these two protagonists, so accustomed to being antagonists.

The phenomenon of plasticity introduces a new dialectic with regard to the organism. Contrary to what seems to be suggested by the conventional idea of genetic determinism, plasticity implies diversity and singularity. Thus psychoanalysis will no longer be able to overshadow the neurosciences and vice versa. Could it be that the subject of psychoanalysis and the subject of the neurosciences are now one and the same? In any case, the phenomenon of plasticity requires us to conceptualize the psychoanalytic subject within the very field of the neurosciences. If the neuronal network is constituted so as to contain the possibility of its modification, if the subject, even as he receives a form, participates in his own formation (Malabou 2000), his own realization—in short, if we accept the notion of plasticity—we must introduce into the field of the neurosciences the question of uniqueness and hence of diversity.

The concept of plasticity challenges the old opposition between an organic etiology and a psychic etiology of mental disturbances. The fact of plasticity shakes up the givens of the equation, to the point where we even come to conceptualize a psychic causality capable of shaping the organic. The same could be said with regard to the relevance of the problem of epigenesis (Changeux 2002) at a time when the Human Genome Project is leading to an increasingly constricted knowledge of genetic determinism. For the level of expression of a given gene can be determined by the particularities of experience, which demonstrates the importance of epigenetic factors in the fulfillment of the genetic program (Kandel 2001a). In the functioning of genes there are in fact mechanisms intended to leave room for experience, and these enter into play in the fulfillment of the genetic program (Cheung and Spielman 2002), as though, when all is said and done, the individual were to appear genetically determined not to be genetically determined.

What is more, plasticity and epigenesis are in league with one another. The usual view is that the influence of experience and the impact of the environment operate between the genotype and its phenotypical expression, that there is an interaction modifying the expression of the genotype (fig. 1.1). Yet we can hold a different view on the basis of the fact of plasticity, which instead leads to the concept of a complex integration between a genetic determination and a psychic or environmental one. The genotype on the one hand and experience or events on the other constitute two heterogeneous dimensions of the different determinisms linked by the phenomenon of plasticity. The notion of plasticity should thus replace that of interaction, for plasticity in fact integrates genome and environment on the same logical level (fig. 1.1).

The model of plasticity sheds new light on the etiology and pathogenesis of mental illnesses, going beyond the reductionism implied by the usual opposition between the organic and the mental. It is quite clear today that we cannot reduce the emergence of psychiatric illnesses to genetic anomalies linked to a single gene, on the model of the monogenetic diseases. For monogenetic diseases

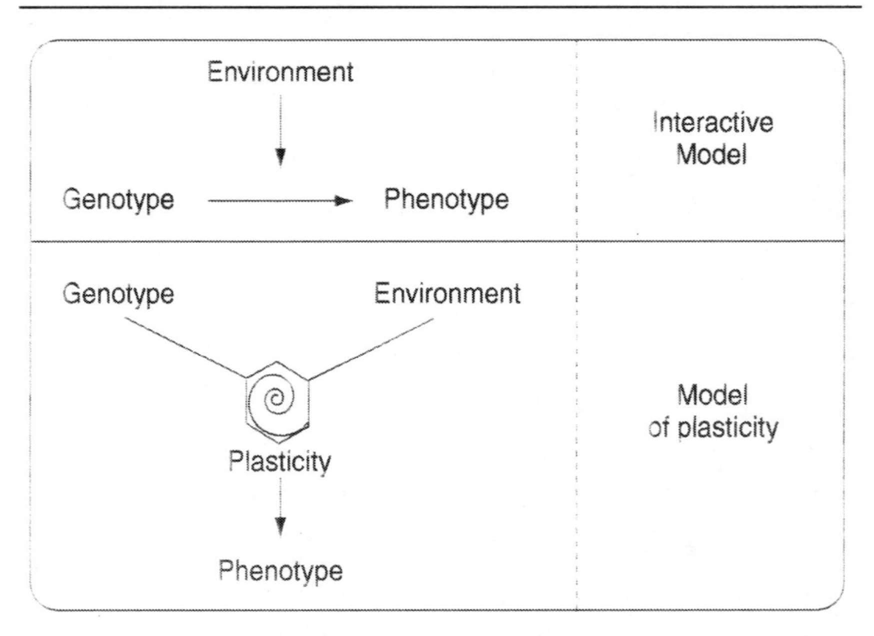

Figure 1.1 The model of plasticity as alternative to the idea of a modulation of the genotype by the environment. In the interactive model, the expression of the genotype is modulated by the environment. In the model of plasticity, genotype and environment constitute two axes of determination that combine via plasticity to produce a unique phenotype.

represent only between 2–3 percent of diseases seen in medical practice, including in psychiatry (Guttmacher and Collins 2003; Schmith et al. 2003).

Nowadays we assume that different genes are implicated in the appearance of a psychiatric illness, more precisely in the likelihood of its appearance (Insel and Collins 2003; Mattay et al. 2003). According to this approach, which is that of the genetics of complex traits, the appearance of an illness thus depends on an interaction between genotype and environment, the characteristics of which are as yet largely unexplained. This approach, though it represents an evolution with regard to a simple genetic determinism, still stops short of the concept of plasticity. Enlightened as it is, it remains within the interactionist model in that it replaces monogenetic determinism with the likelihood of plurigenetic origin.

What makes the concept of plasticity possible is a critical approach to the modification of the expression of the genotype by environmental factors beyond the idea of interaction. What we have here are two determinisms, parallel but different: a genetic determinism, plurigenetic to be sure, and an environmental or mental determinism, which are linked in a special way in the phenomenon of plasticity. In this view there would no longer be genetic determination any more than there would be environmental or mental determination. On the contrary, there would be two determinations whose connection is to be conceptualized through the phenomenon of plasticity.

Plasticity thus enables us to take maximal advantage of the spectrum of possible differences, leaving due place to the unpredictable in the construction of individuality, and the individual can be considered to be biologically determined to be free, that is, to constitute an exception to the universal that carries him.

Hence plasticity entails moving on to a new paradigm and enables us to effect a scientific revolution in the Kuhnian sense (Kuhn 1970). For Kuhn, when a paradigm is pushed to its extreme point—for example, the paradigm of the organic determination of the psyche, or indeed the paradigm of the genetic determination (Atlan 1999) of human behavior—it becomes exhausted until it ends in failure, thereby opening the path toward a new conception. This crucial stage cannot be skipped. Psychoanalysis and the neurosciences should instruct one another reciprocally, starting out from the stumbling blocks in their specific domains and daring to explore what resists their progress. For psychoanalysis, this would be to go in the direction opened out by Lacan (1964) when he asked what a science would be like that included psychoanalysis. For the neurosciences, this would be to find in psychoanalysis the necessary fulcrums for orienting themselves to the emergence of the unique within the general biological mechanisms it reveals.

Thus we suggest that psychoanalysis be linked to the neurosciences by the concept of plasticity. Though it comes straight from biology, this concept proves operative in the field of psychoanalysis. The incommensurability of these two fields, however, remains a stubborn fact. There is no syncretism between the neurosciences and psychoanalysis, no reconciliation, no possible synthesis. There is no salvation for thought unless we first acknowledge the essential differences that exist between the neurosciences and psychoanalysis. These differences are a dynamic factor from which the emergence of the subject can be inferred, including on the basis of the laws of biology.

How, then, shall we conceptualize the relation between the neurosciences and psychoanalysis from the fact of plasticity? Right from the start plasticity eliminates the idea of an absolute heterogeneity as well as that of a superposition without distinction (fig. 1.2). To say that the neurosciences and psychoanalysis belong to two heterogeneous orders does not mean that they are without any relation. The phenomenon of plasticity itself argues against such a view. We could even formulate the paradoxical hypothesis that it is because they are incommensurate that the neurosciences and psychoanalysis can be connected. It remains to be seen how two heterogeneous orders can be linked. Is there a merging to form a unity, or is there only an intersection between two heterogeneous orders in which one affects the other and vice versa (fig. 1.2)?

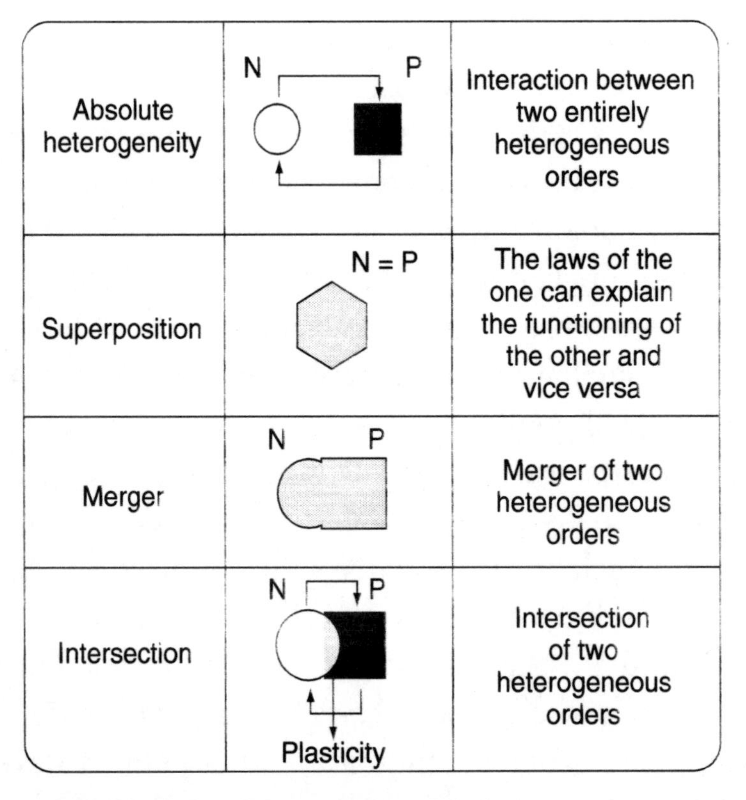

Figure 1.2 Models of the relationship between neuroscience and psychoanalysis

Merger and intersection are associated with a difference in terminology. The merger model implies a denaturation of the characteristics proper to each order, which would lead to mixing the neurosciences and psychoanalysis into an undifferentiated whole. The intersection model, in contrast, accepts that the neurosciences and psychoanalysis can come together on account of plasticity even as the existence of elements having nothing in common is respected. Only the latter model takes into account the phenomenon of plasticity, in which the mental marks the organic, having an effect on matter by leaving concrete material traces as functions of experience. What plasticity demonstrates is, in fact, that the neurosciences and psychoanalysis, domains whose dimensions are incommensurate, can mutually affect one another.[3]

---

3. The concept of plasticity means that experience can be inscribed in the neuronal network. An event experienced at a given time is marked at the moment and can persist over time. The event leaves a trace and, simultaneously, time is embodied. But this trace can be reworked or put in play again in a different way by being associated with different traces. Beyond biological determinism (neuronal or genetic), and beyond psychic determinism, the fact of plasticity thus involves a subject who actively participates in the process of his or her becoming, including that of his or her neuronal network. Hence we discover in a new form the relevance of Lacan's (1959–1960) emphasis on the importance of understanding how the organism comes to be caught up in the dialectic of the subject.

The area of intersection of these two heterogeneous orders raises the question of the trace left by experience through the mechanisms of plasticity. The question of the trace is common to the two heterogeneous orders. The close relation linking the synaptic trace and the mental trace is thus at the heart of this book.

2

Diego and Haydn: Perception and Memories

Imagine yourself on Christmas night, seated at the family table, admiring the decorated pine tree, savoring a turkey accompanied by a delicious 1990 Barolo. In the background the notes of a Haydn concerto are covered by affectionate conversations suited to the occasion. You are patting the soft fur of Diego, your Labrador retriever, who has come to beg for his unlikely part of the banquet. Your brain perceives all these pieces of information almost instantaneously via the different sensory modalities—touch, sight, hearing, smell, and taste—that make possible the perception of stimuli coming from the external world. The nervous impulses gallop along your nerves at a speed of around two hundred miles an hour. From the retina, the tympanum, the skin, the tongue, and the mucous membranes of the nose, these highways of sensations that are the nervous fibers of the sensory systems convey to the brain in several tenths of a second the pieces of information from the external world. And there are many others: the decorations on the tree, the texture and pattern of the tablecloth, the taste and smells of the other dishes. Perceptions flood your brain.

But something else is happening as well. This Christmas night calls to mind many others from the past, happy or sad, and multiple memories of this year's party will remain anchored in your memory. Here we find a second component of brain functioning: perception can leave a trace in the nervous system and become memory. In other words, perception leaves a sign inscribed in the neural circuits, one that could be identified with the Freudian concept of the sign of perception.[1]

Recent advances in neurobiology have made it possible to clarify some of the molecular and cellular mechanisms effecting the inscription of this trace, namely the constitution of a memory. How does perception leave a trace, that is, how can lived experience be inscribed in the networks of neurons? The modalities of this inscription, hence the mechanisms of memory, are based on an essential property of the nervous system: neuronal plasticity. What is this exactly?

The brain is sometimes seen as a system functioning in binary fashion, with information either passing or not passing in the circuits, as if basic elements, the neurons, were organized like microcircuits engraved in sili-

1. Freud 1887–1902, 174. See also chapter 5, "Forgetting the Name *Signorelli.*"

con, like those of a computer. Such a view, relatively simplistic and rigid, does not correspond to recently obtained experimental data, according to which information is in fact transmitted in our brain from one neuron to another in a highly modulated manner. As an initial analogy, let us consider a rheostat, the two poles, that is, of a binary communication. The neuronal circuits have very little to do with the microcircuits engraved once and forever at the end of an assembly line (in a biological context we would say at the end of the development of the nervous system). For what we find here is a concept mentioned above, that of plasticity. Plasticity is the reverse of rigidity. For neuronal circuits it involves the ability of the neurons to modify the efficiency with which they transmit information (Bear 2003).

What property, then, does this neuronal plasticity provide for our brain? It enables the brain to register in a lasting way pieces of information coming from our environment, making it possible for the experiences undergone by each individual to leave a trace in the neuronal circuits. As we shall see, the term *trace* has not been usurped, even in the biological sense, for these are molecular and cellular traces left on the level of the subtlest mechanisms of neuronal functioning.

These mechanisms of neuronal plasticity have been most closely studied in the context of the processes of learning and memory, but it is legitimate to suppose that they may be extended to all of an individual's experiences, especially to what contemporary neuroscientists call emotional memory (LeDoux 1996).

Let us now try to go from concepts to matter, the matter of which the brain is made. The transition is not easy, for here we find the previously mentioned factors of incommensurability between psychoanalysis and the neurosciences. Yet it is absolutely necessary to deal with it. For plasticity is not only a concept; it is a biological reality that gives rise to the notion of the subject's uniqueness.

Current estimates put the number of neurons constituting the brain at over one hundred billion (Bear, Connors, and Paradiso 2001). They exist in different "models," these mysterious butterflies of the mind, as the Spanish neurobiologist Santiago Ramon y Cajal calls them.[2] Depending on their form, they correspond to evocative names: the double-bouquet, chandelier, stellate, pyramidal, and bipolar neurons, and Purkinje's neuron, named after the Czech histologist who described them in the nineteenth century, to mention only a few.

Despite their diversity of shape and size—certain neurons, like the motor neurons controlling the muscles of the toes, have processes over a meter long, going from the spinal cord to the foot, while others project only a few fractions of a millimeter—their functioning is relatively uniform. This is fortunate for the neurobiologist,

2. "Like the entomologist seeking colorful butterflies, my attention has hunted delicately and elegantly formed cells in the orchard of gray matter, the mysterious butterflies of the soul, whose wing beats may one day yield up the secrets of the mind to us" (Ramon y Cajal 1909–1911, 98–99).

for how could he hope to take an interest in the brain if each of the hundred billion neurons obeyed separate laws of functioning? But, luckily, universal mechanisms can coexist with morphological diversity (for example, the shapes of neurons) and functional uniqueness (for example, the function of each neuron in a circuit).

From the functional perspective each neuron has three parts: a receptive area, the dendrite, which receives information from other neurons; an area, the cell body, which integrates the received information; and the axon, a part through which it sends signals for the other neurons. The mechanisms of plasticity that concern us here are centered on the contacts among the neurons, the place where they exchange information. This zone of contact among the neurons is the synapse. It includes a presynaptic part, located at the terminal of the axon, and a postsynaptic part, generally corresponding to a specialized area of the dendrite called the dendritic spine, resembling the thorn on a rose stem (fig. 2.1)

Each neuron receives around ten thousand synapses from other neurons, which gives a million billion points of contact to which information can be transmitted among neurons. These are dizzying numbers, even more so because the efficiency with which information is transmitted from one neuron to another, at each of these contact points (the synapses), varies in the course of life depending on experience. We are very far indeed from the notion of rigid, binary wiring.

The presynaptic part of the synapse contains small sac-like objects called vesicles, inside of which thousands

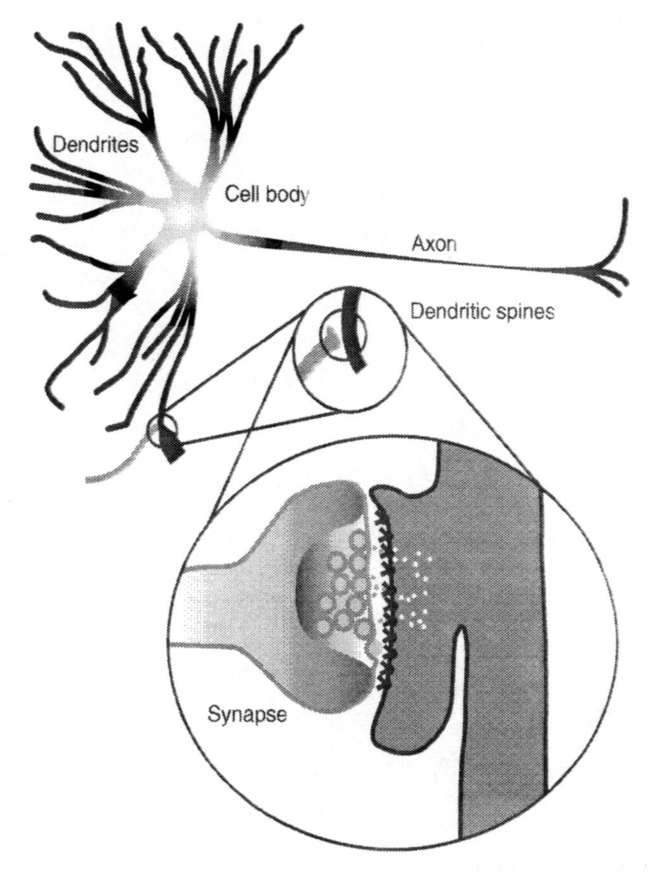

Figure 2.1 Structure of the neuron

of molecules, the neurotransmitters, are collected. Each neuron contains a primary neurotransmitter and sometimes one or more (rarely more than three) so-called accessory neurotransmitters. The neurotransmitters are the molecules by means of which the neurons send their signals. They are released when the axon terminal is activated and the vesicles, merging with the presynaptic membrane by highly regulated mechanisms, pour their contents of neurotransmitters into the synaptic gap, that tiny space, several millionths of a millimeter wide, separating the pre- and postsynaptic sides (fig. 2.2).

This quick overview of the process enables us to glimpse an initial possibility of modification, that is, of plasticity, with regard to the transfer of information among neurons, since variable quantities of neurotransmitters can be released. For a neuron can be influenced, on a long-term basis, to release more neurotransmitters for one and the same situation (Bliss, Collingridge, and Morris 2003). Depending on the degree of activation of the presynaptic terminal, a varying number of vesicles will merge with the presynaptic membrane in a unit of

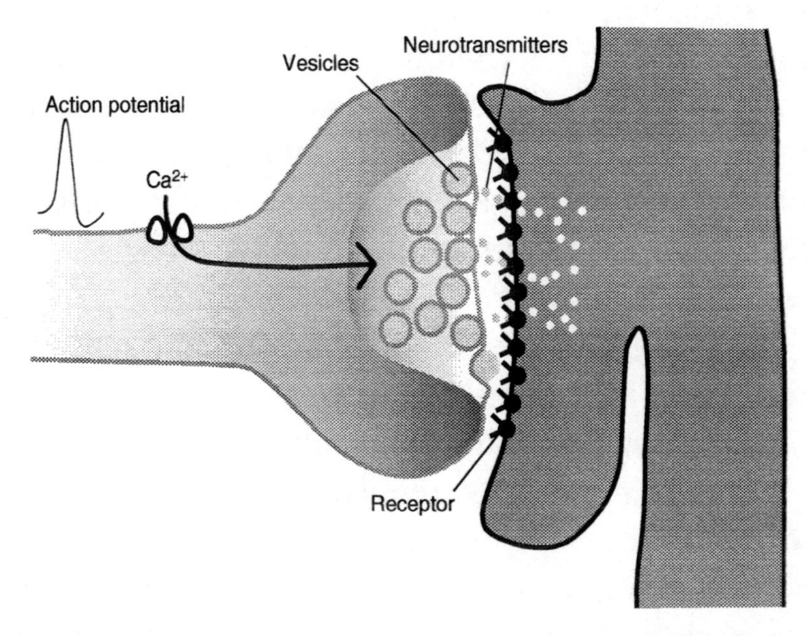

Figure 2.2 Release of neurotransmitters from the presynaptic terminal. Neurotransmitters act on the receptors on the postsynaptic side.

time. To the extent that the number of neurotransmitter molecules per vesicle is relatively uniform (several tens of thousands), for a given presynaptic activation a given number of neurotransmitter molecules will be in the synaptic cleft and will trigger a given response on the postsynaptic level. By way of a loose analogy, we can speak of a kind of presynaptic rheostat, like the kind that enables us to vary the intensity of light in a room.

The mechanisms activating the presynaptic terminal, resulting in the release of neurotransmitter molecules, have been described in detail. As with every cell of our organism, there is a difference in electrical potential between the inside and the outside of the neuron. This difference is weak, on the order of sixty to ninety thousandths of a volt (millivolts, mV), but it is enough to generate currents. To understand this better, let us think of the two poles of an electrical battery that make it possible to produce a current between the positive pole and the negative pole, with a difference in voltage on the order of several volts.

By convention, the inside of the neuron is negative with respect to the outside. This difference in potential is due, among other things, to an unequal distribution of ions (charged atoms) present on both sides of the cell membrane, the extracellular region being rich in sodium and calcium whereas the cytoplasm is rich in potassium. The passage of ions across the membrane of the neurons generates currents. When a neuron is activated, brief currents due to the passage of sodium (thus of positive charges) from the outside to the inside of the neuron (these currents last around five thousandths of a second) are produced along the axon, which temporarily makes the inside of the neuron positive. Thus the potential across the membrane goes from a very negative value, for example, –70 mV, to a very positive one, generally +60 mV. We then say that the neuronal membrane is depolarized. This temporary change of the potential of the membrane, which is on the order of 130 mV and propagated along the axon, is called the action potential (fig. 2.3, lower panel).

This action potential is produced at the junction between the cell body and the initial segment of the axon. As it is propagated, it penetrates the axon terminal and facilitates the entry of calcium there.

The resulting increase in the concentration of calcium is the signal triggering the fusion of the vesicles with the presynaptic membrane. Generally speaking, the higher the concentration of calcium, the more likely it is for the vesicles to fuse with the presynaptic membrane; hence more neurotransmitter molecules are released into the synaptic cleft. Here we see another possible level of plasticity: every process that durably modifies the concentration of calcium obtained by the activation of the presynaptic terminal influences the amount of the neurotransmitter that is released.

Let us move on to the second element of the synapse, its postsynaptic side, generally located on the dendritic spines of the neuron receiving the information. What happens here, once the neurotransmitter molecules have been released? These molecules are recognized in a highly

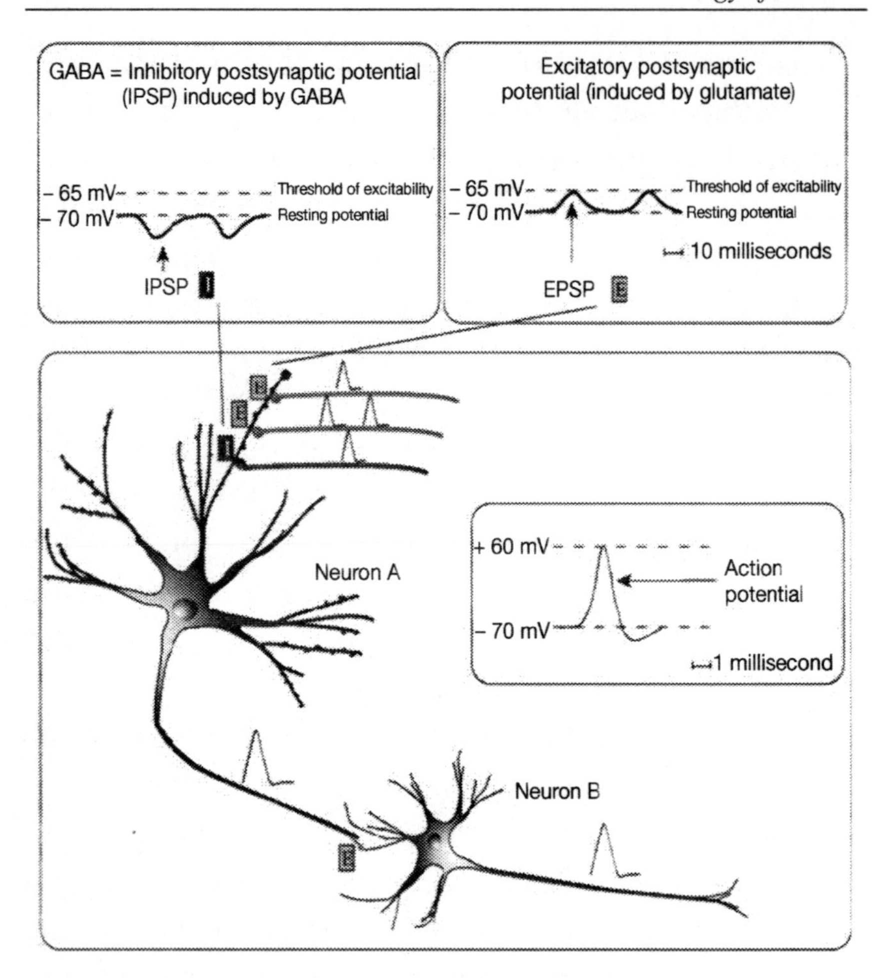

Figure 2.3 Mechanisms of signaling between neurons. Neurons communicate with each other through electrical signals (action potentials) and chemical signals (neurotransmitters, for example glutamate or GABA). The latter make the target neuron (Neuron B in the diagram) more excitable (EPSP) or less excitable (IPSP).

specific manner by receptors on the postsynaptic membrane (an analogy is often made to a key and a lock for this kind of interaction) (fig. 2.2). The receptors are complex molecules that, as it were, float in the neuronal membrane. Some of them face the synaptic cleft and recognize the neurotransmitter, while others cross the thickness of the membrane (less than one millionth of a millimeter) and span to the inside of the cell, entering into contact with the intracellular region called the cytoplasm.

When a neurotransmitter activates a receptor called ionotropic, one of the two broad categories of receptors, the ionotropic receptor changes its conformation and creates a channel (similar to those described in the case of the action potential). For several thousandths of a second, this channel establishes communication between the extracellular region and the interior of the cell. This temporary opening allows for the creation of an electric current, charges carried by ions present in the extracellular region—sodium and/or calcium (positive charges) or chlorine (negative)—that make the potential of the postsynaptic neuron more positive or negative, respectively.

In this way, the neurotransmitters acting on the ionotropic receptors have the power to make a neuron either more excitable, in which case the transfer of information will be facilitated, or less excitable, removing this neuron from the circuit in a certain way (fig. 2.3). These influences on excitability can be the site of a plasticity that lastingly enhances the physiological action of a neurotransmitter.

Two principal neurotransmitters effect the transfer of information to the synapses of the nervous system: glutamate, which increases neuronal excitability, and gamma-aminobutyric acid (GABA), which diminishes this excitability.

The response leading to depolarization is called excitatory postsynaptic potential (EPSP), and the hyperpolarizing response is called inhibitory postsynaptic potential (IPSP). In contrast to the action potentials having an amplitude of 100 to 150 mV, the synaptic potentials are of small amplitude, on the order of a few mV. These two primary neurotransmitters induce the following responses: glutamate for the EPSP and GABA for the IPSP. It is important to note that the synapses releasing either glutamate or GABA represent over 90 percent of the synapses of the nervous system (fig. 2.3).

The neurons integrate the EPSP and the IPSP they receive (each neuron has up to ten thousand synaptic contacts, some of which are active simultaneously) according to relatively complex and subtle mechanisms. But in order to simplify we can say that, if the EPSPs predominate, a neuron A (fig. 2.3) will be depolarized and action potentials will propagate along its axon. This, in turn, will prompt the release of neurotransmitters at its axon terminal, triggering a response (EPSP, IPSP) in neuron B.

If, for example, neuron A releases glutamate and produces an EPSP in neuron B (fig. 2.3), the latter will in turn integrate these synaptic responses. If, in contrast, the IPSPs predominate, neuron B will be hyperpolarized, thereby reducing the likelihood that action potentials will be generated in its axon: the neuron will be inhibited, will be somehow "out of the circuit." To this we may add that the mechanisms of synaptic integration are such that, when several synapses (for example, excitatory ones) are simultaneously active on a neuron, the EPSPs generated at each synapse add up so as to produce an EPSP of several tens of mV. This phenomenon is called spatial summation (fig. 2.4).

In addition to the ionotropic receptors there is a second type of receptor, the metabotropic. These receptors recognize neurotransmitters just as the ionotropics do. Following their interaction with the neurotransmitters, however, they do not create a channel but activate enzymes

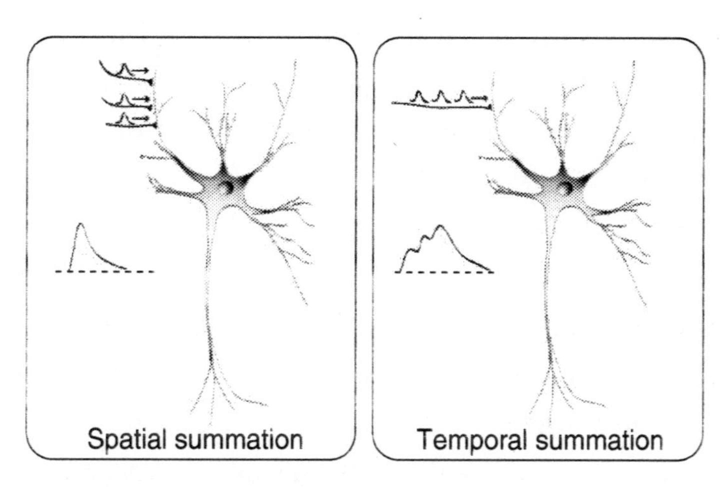

Figure 2.4 Synaptic integration.
Excitatory postsynaptic potentials (EPSPs) generated at several synapses are added together (spatial summation). When several postsynaptic potentials are generated in a limited time period (several tenths of a second) in a single synapse, they too are added together (temporal summation).

present in the membrane. These enzymes are responsible for forming new molecules called second messengers in contrast to the neurotransmitters, which are the primary messengers of communication between neurons.

These second messengers control various cellular processes that modify the activity of postsynaptic neurons. Two of these cellular processes modified by the second messengers are especially relevant for the analysis of the transfer of information between neurons and their plasticity.

On the one hand, certain second messengers modify the activity of ionotropic receptors, prolonging by several milliseconds the time of the opening of the channel created by the ionotropic receptor. This leads to an increase in the number of ions (and hence of charges) passing across the channel, which in turn increases the effectiveness of the neurotransmitter. On the other hand, other second messengers mobilize ionotropic receptors that are, as it were, held in reserve near the membrane and bring about their insertion into the membrane. The membrane of the postsynaptic neuron will then be richer in ionotropic receptors and will respond more effectively to the neurotransmitter released by the presynaptic terminal; in the case of an excitatory neurotransmitter its effect on neuronal excitability will be increased.

These mechanisms can operate on a long-term basis, hence lastingly modifying the transfer of information at a synapse. Here we have further examples demonstrating the possibility of plasticity for physiological mechanisms.

If a synapse is the site of information transfer between neurons, this transfer is thus not binary in nature or constant in its intensity; on the contrary, it is highly modifiable. Experimental research has revealed a regulation in three stages. The first is located at the presynaptic level, the two others at the postsynaptic level. They concern the release of neurotransmitters from the axon terminal (this mechanism depends on the concentration of calcium: the higher the concentration, the greater the number of neurotransmitter molecules released), and also the activity and density of the ionotropic postsynaptic receptors, which are modified by the second messengers. These regulatory mechanisms make it possible to establish conditions of plasticity through the durable modifications of synaptic efficiency associated with the processes of learning and memory and thus with the constitution of a trace in the neuronal network (Bliss, Collingridge, and Morris 2003).

The brain, then, has mechanisms allowing for the perception of the external world and other mechanisms, whose elements we have outlined, allowing for the inscription of these perceptions in the neuronal network and the formation of memories. We have outlined the molecular mechanisms that, by modifying the intensity of the transfer of information between neurons at the level of the synapses, contribute to the establishment of a trace. The mechanisms by which a set of traces can constitute a representation, for example a memory, will be discussed in chapter 5.

It is clear that the purpose of perceptions is not solely to enhance memory or end in learning; they trigger motor responses that, fortunately for us, are adequate most of the time. We all know what to do with a Christmas turkey: carve it and savor it. This is obviously not a reflexive motor response like the one that leads to the extension of the leg when the doctor strikes with his hammer the quadriceps tendon just below the kneecap; the mechanisms of synaptic plasticity have resulted in a learned motor response that makes it possible for us to implement the operations necessary for feasting on the turkey. And this does not happen any old way: the motor act is modified by the cultural context, that is, by the good table manners we have been taught. We do not behave like a Neanderthal man before a leg of raw boar meat; we skillfully carve the turkey thigh, handling the utensils according to a well-established code.

Thus the brain has fine mechanisms for storing perceptions and recalling them when necessary, sometimes in a way we could call automatic, as in the case of motor learning. In this case we could say that this is a *non-conscious type of memory*, which some call procedural memory (Eichenbaum et al. 1999): it is not necessary to consciously review the different movements that enable us to eat a turkey thigh in an elegant fashion, for we do so automatically. But if we have to explain the steps, for example when, in our role as parents, we have to teach our child good table manners, we can recall the operation and evoke it with great precision down to the smallest detail. In the same way a golf pro will hit the ball on the course without thinking about it, but when teaching he will describe to his student with infinite exactitude the slightest nuances of the swing.

We have deliberately used the term *non-conscious* here in place of *unconscious*. In the literature exploring the frontiers and the correspondences between the neurosciences and psychoanalysis the non-conscious and procedural memory have too often been equated with the unconscious. In our opinion the term *unconscious* is to be understood in the Freudian sense.[3] Freud's idea of the unconscious is associated with the notion of a series of entirely unique traces and associations that are not immediately accessible to consciousness except through dreams, slips of the tongue, lapses of memory, bungled acts, and other formations of the unconscious whose meanings can be uncovered in psychoanalytic work.[4]

3. "The unconscious is the larger sphere, which includes within it the smaller sphere of the conscious. Everything conscious has an unconscious preliminary state; whereas what is unconscious may remain at that stage and nevertheless claim to be regarded as having the full value of a psychical process. The unconscious is the true psychical reality; *in its innermost nature it is as much unknown to us as the reality of the external world, and it is as incompletely presented by the data of consciousness as is the external world by the communications of our sense organs.*" (1900, 612–613, emphasis in original).

4. Compare Lacan's statement that the phenomena emerging from the unconscious—the formations of the unconscious—are structured like a language (see Lacan 1966). In this view the unconscious would then be closer to the processes of declarative memory than to those of procedural memory, with the difference that, in contrast to the memories of events and objects of external reality

Hence we find it all the more unjustified to equate the unconscious with procedural memory insofar as experiences registered in a procedural mode can be very easily recalled to awareness, as in the case of the golf pro.

At first glance, then, the memory systems allow access to experience that has been registered in the form of learning or memory in a way that preserves a remarkable correspondence with what was initially perceived. Through the mechanisms of synaptic plasticity, which make possible the establishment of a trace in the neuronal network on the basis of perception of the external world, an internal reality is constituted, one that we are aware of or that can emerge into awareness through recall.

But things are not so simple. Something else is happening on this Christmas night. Suddenly you are overcome by a great sadness. You are seized by a feeling of devaluation that becomes more and more insistent to the point of disgust. You don't understand: everything is there for your happiness, the good cheer, the gifts, Diego, the music, but none of this remains. It all turns into the opposite. Another associative chain has now interfered with the present situation in connection with a recent business deal that turned out badly, leaving you with the impression that you were cheated. Your gaze becomes unfocused as you mechanically carve the turkey. An idea crosses your mind and imposes a series of associations

---

that are directly accessible to consciousness via simple recall, unconscious declarative memory would involve passage through processes of association facilitated by psychoanalytic work.

that go from the turkey on the table to the inept turkey you were in the business deal.

In this way an ordinary event in the present situation is juxtaposed with something else, and you find yourself in a mental world that no longer has anything to do with that present situation. There is no longer any correspondence between the present perception, the Christmas turkey, and the memory of the fact of having been cheated. A current perception gives rise to an entirely different representation coming from the internal world. Through the mechanisms of conscious declarative memory, the same perceived object at first evoked recollections corresponding to the reality of the situation (for example, the Christmas celebrations of your childhood), but, in an associative slippage, this same object, the turkey, activated a painful situation you thought you had set aside in the present circumstances.

So there you are, caught in a conflict between the present and the past. But this conflict also makes it possible for you to disengage from it, to go back to where you were, and, by becoming aware of it, to separate yourself from this constraining sequence of associations. Here we find the idea of a possible modification of the trace, in that it can be modified by its very explanation. Once again, this is what can facilitate analytic work, which counts on the effects of associations among different traces so as to modify their expression.

3

Inhibition on the Shore of Lake Trasimene: What Becomes of Perception

In August 1897 Freud (1887–1902)* described to Wilhelm Fliess, his principal correspondent at the time, his tormenting doubts about his theory of the neuroses. Two years had passed since the publication, with his colleague Joseph Breuer, of the *Studies on Hysteria* (Breuer and Freud 1893–1895), a work that paved the way for psychoanalysis by assigning a traumatic etiology to this neurosis, most often the result of a seduction by a relative or close friend that took place in reality; its recollection was now kept out of awareness, repressed, even as it remained active deep inside, playing a role in symptom formation unbeknownst to the patient. The symptoms thus took the place of what the subject could not remember.

*All references to the Freud–Fliess correspondence are to this volume.

Freud counted on speech—the "talking cure," as it was called by Anna O., a young hysteric who became his patient after being treated by Breuer—to bring the trauma into awareness and facilitate what he termed an abreaction, that is, an affective discharge through which the subject could be freed from the distress accumulated under the pressure of the forgotten event. Analytic work would bring the trauma to light, making it conscious and thereby releasing the patient from its constricting effects.

Such a concept remains within the cathartic approach to treatment, on the model of hypnosis: all one had to do in order to undo the effects of the trauma was to remove the amnesia, bringing the event into awareness. Yet, as we know, despite the effective nature of this hypothesis, the treatments did not turn out to be as effective as Freud had anticipated, and he had to deal with repeated disappointments.

This is what tormented him in this month of August 1897. As he wrote to Fliess, "Things are fermenting inside me, but I have nothing ready." Seized with doubts, he was looking for a solution. But he came up against something that paralyzed his work, namely the feeling that it was within himself that the solution was to be found. He therefore became his own patient: "The chief patient I am busy with is myself." It was in this context of gloom that he decided to take a step he knew would be right for him: he prescribed a journey for himself: "I am . . . very lazy and have done nothing here to get the better of the turbulence of my thoughts and feelings. That must wait for Italy" (p. 213). The itinerary was set. He would begin in Venice then move on to, among other places, San Giminiano, Sienna, Perugia, and Assisi, in other words Tuscany and Umbria. Freud kept up a steady pace on this trip, his intention being to end up in Rome. But then, on the shore of Lake Trasimene, he was overcome by an inhibition and could not continue. He changed his plan and suddenly decided to return to Vienna.

The day after he got back he wrote Fliess a letter, destined to become famous in the history of psychoanalysis, in which he announced that he no longer believed in his "*neurotica*" (p. 215). A major step had been taken. For Freud was now subjecting to a radical critique the positions he had taken up to that time regarding the etiology of the neuroses. It was the very hypothesis of seduction that he had now rejected. An event does not have to occur in reality in order to bring on a neurosis: an imaginary construction produced by the subject himself can suffice.

Thus the origin of the neuroses has shifted from the traumatic reality of seduction to a seduction fantasy linked to wishes that exert pressure even as they remain out of awareness. Internal stimuli can mark psychic life, beyond all reality, in the absence of an actual traumatic event. A fantasy can be enough to organize the symptoms of a neurosis. Psychic reality takes precedence over external reality in Freudian thought, and so there is no point in making an exhaustive search for the causal event, especially since, he tells Fliess, there is no "'indication of reality'" in the unconscious, even though it is on this level that the earliest experiences are inscribed and "it is impossible to distinguish between truth and emotionally-charged fiction" (p. 216). In hunting for the event we may come upon a fantasy that was constituted in obedience to other laws than those of reality: the laws of unconscious wishes.

As Lacan (1967) put it, fantasy provides the context for reality. It takes part in constituting reality as understood by the subject. Thus it is useless to search for the forgotten memory, for this memory is included in the fantasy, disguised by all sorts of psychic mechanisms similar to those at work in dreams.[1] "Even in the most deep-reaching psychoses the unconscious memory does not break through" (Freud 1887–1902, 216). The experienced event does not present itself. Yet this is not a reason to stop trying to go in search of it in the form in which it is revealed under the pressure of the unconscious wish. And this is what Freud attempted on his return from Lake Trasimene, undertaking very actively what he described as his self-analysis. As we know, this analysis would set him on Oedipus' road in terms of his own biography.

What, then, happened on the shore of Lake Trasimene; what is the source of the inhibition that came over him fifty miles from Rome? He tells us in *The Interpretation of Dreams*: "It was on my last journey to Italy, which, among other places, took me past Lake Trasimene, that

---

1. "It is true that as a rule the childhood scene is only represented in the dream's manifest content by an allusion and has to be arrived at by an interpretation of the dream" (Freud 1900, 199).

finally—after having seen the Tiber and sadly turned back when I was only fifty miles from Rome—I discovered the way in which my longing for the eternal city had been reinforced by impressions from my youth" (1900, 196). What Freud remembered was Hannibal, a favorite personage from his childhood. For this Carthaginian, a Semitic hero, was associated with an event in Freud's early life.

He was walking one day with his father, who described to him a humiliating experience he, the father, had had, one that bore witness to the anti-Semitic climate of that time. The father had gone out into the street well dressed, wearing a brand-new fur cap. A Christian knocked the cap into the mud, crying, "'Jew! Get off the pavement!'" "'And what did you do?'" asked the young Freud. The father told how he complied with resignation, stepping off the sidewalk to retrieve his cap. This outcome left its mark on little Sigmund: how could this "big, strong man who was holding the little boy by the hand" let himself be treated this way? This became a memory of the father's demotion from a position of prestige, and Freud "contrasted this situation with another which fitted my feelings better: the scene in which Hannibal's father, Hamilcar Barca, made his boy swear . . . to take vengeance on the Romans. Ever since that time Hannibal had had a place in my phantasies" (1900, 197).

Now at the time of the journey through Tuscany and Umbria, Freud was in the process of applying for a professorship, but the academic authorities in Vienna at that time were anti-Semitic, and he had to take steps to support his candidacy, which, for him, represented a compromise of principle somewhat like his father's in the cap episode. The fantasy of identification with Hannibal was certainly in play on the unconscious level at the moment when Freud, like his childhood hero, found himself on the shore of Lake Trasimene in September 1897. Like Hannibal, he did not allow himself to go all the way to Rome, did not give himself the right to avenge his father, that is, likewise to go beyond what his father was able to do and thereby surpass him. Vanquished unconsciously, he turned around and returned to Vienna, confirming there the extent to which the unconscious determination of mental life can take precedence over real events. This, at least, is what he came to understand later on, as he revisited this inhibition in his self-analysis, which would eventually make clear the weight of unconscious feelings of love for his mother and rivalry with his father, strikingly revealed to him in King Oedipus and Hamlet.

Our concern here is not to retrace the path of this discovery in Freud's self-analysis but rather to make use of these developments to show the extent to which lived experience can get lost and transformed in the defiles of its mental inscription. Though experience does leave a trace, it can be reinscribed several times over in a different way, and it is thus, on the basis of one of its later fates, that at a particular moment it can become determinative for the subject. Freud's anecdote leads us to distinguish reality and mental reality, questioning the relation—or why not also the non-relation?—between experience and the trace it leaves on the neuronal level and, in addition, its mental effect.

Relation or non-relation? As the story of the inhibition on the shore of Lake Trasimene shows, this is a complex question: the experience gets lost in the associations it begets through the very mechanisms of its inscription. The trace of the experience inscribed through the mechanisms of plasticity can undergo many reworkings and become associated with other traces, distancing the subject from the event that took place. These mechanisms of association operate in such a way that mental reality goes beyond the experiences that caused the initial trace.

To put it in other words, a set of traces that are associated and combined substitute for the experience. The system gets complicated to the point where it becomes organized in the form of new stimuli: a mental reality prevails over external reality, which from then on becomes, as Freud puts it, fundamentally "unknowable" (1938, 196).

Here we have a paradox. The mechanisms permitting the inscription of the experience are those that separate us from the experience. We find a trace, but we no longer find the experience, all the more so because this trace is recombined with other traces according to new laws proper to mental life. Even if, as Freud says, there is perception at the outset, when it is inscribed it becomes a stimulus of another order for the neuronal apparatus. Thus from transcription to transcription the experience as such gets lost by means of the mechanisms of synaptic plasticity, even though it has produced durable traces.

Thus there seems to be a sort of contradiction in psychoanalytic theory with regard to perception. On the one hand, as we have already noted, for Freud "all presentations issue from perceptions" (1925a, 237). But, on the other hand, the processes of mental life and inscription make it "impossible to trace their original connection" (1887–1902, 204). As a result of the entire series of combinations, the experience becomes inaccessible as such.

Through the processes of association, fusion, deformation, modification, and fragmentation the experience is reinscribed several times. It takes a new form, for example that of a fantasy. As Freud writes, fantasies are "constructed by a process of fusion and distortion" (1887–1902, 204), ending in a falsification of the scene that occurs, the chronological considerations notwithstanding, beyond the experience that really took place, with certain unused fragments of reality also able to enter into this combination. From one inscription to the other, from trace to trace, we no longer find the lived experience but instead a series of fantasies that, from now on, will determine mental life as such.

Though some scenes remain accessible, most will be repressed and modified by the intermediation of "superimposed" fantasies. "Those which are more slightly repressed come to light only incompletely . . . on account of their association with those which are severely repressed" (1887–1902, 202–203), all the while getting confused by fantasies that will be formed on this basis. In short, everything will be interfered with by fantasy, which, Freud says, combines lived incidents, accounts of past facts, and things seen by the subject himself. Though the experience and the fantasy remain somehow connected, the fantasy has become a new source for mental life. The circumstances of the event are caught in the defiles of its inscription so as to constitute a fantasy. The connection with the event is now inaccessible; we can no longer go back from the fantasy to the event.

In our opinion the inscription, transcription, and association of traces left by experience are effected by the mechanisms of synaptic plasticity. The famous experience that leaves a trace is no longer solely the external event but also the event inscribed and transcribed by the mechanisms of plasticity. We are very far from the perceived event. This is also what Freud states both in his investigation into the etiology of hysteria and in his self-analysis: "The secret of infantile experiences is not revealed" (1897–1902, 216). In its place a whole series of associations of another order enters into play, from the father's cap to Hannibal, passing via the professor's dossier all the way to Lake Trasimene in our anecdote. We can see how impossible it has become in mental life to reestablish a direct link with perception. The etiological role of the event can only be put back in question.

We understand that Freud had to renounce his *neurotica.* Instead of seeking a factual etiology for the neuroses he became interested in the world of fantasy, that "half-way region" (1912, 252) in which experience is retranscribed in a new way. In his view fantasy feeds awareness just as perceptions do, determining the subject's actions and mental productions.[2] In this way part of mental reality is separate from external reality, remaining independent of it and obeying different laws. The creation of fantasies ends in a new excitation of the neuronal apparatus that takes the place of external excitation (Freud 1911). The incidental event may thus be internal, a sort of intrapsychic perception that must be taken into account when we want to construe what a psychic event is, that famous experience with which we began and that has marked with its trace the organization of the neuronal network.

2. He writes, for example, "All anxiety symptoms (phobias) are derived in this way from phantasies" (1887–1902, 204).

4

Aplysia, Rat, Man: From Experience to the Trace

There is a simple organism, *Aplysia californica*, a sea snail that wades in the waters of California, to which we owe a great deal with regard to the understanding of the molecular mechanisms of synaptic plasticity linked with the processes of memory and learning (fig. 4.1.A).

The groundwork on synaptic plasticity in this simple biological specimen was laid forty years ago (Kandel 2001a, 1030–1038), but it provides an excellent illustration of the essential principles by which learning, even simple learning, leaves a trace in the synapses. It is a bit as though we were learning the alphabet of synaptic plasticity to which we have been constantly referring.

The nervous system of Aplysia has about a thousand large neurons, a windfall for neurobiologists who can easily implant electrodes there. Despite its limited number of neurons and a limited behavioral repertoire, Aplysia is capable of certain simple, quantifiable forms

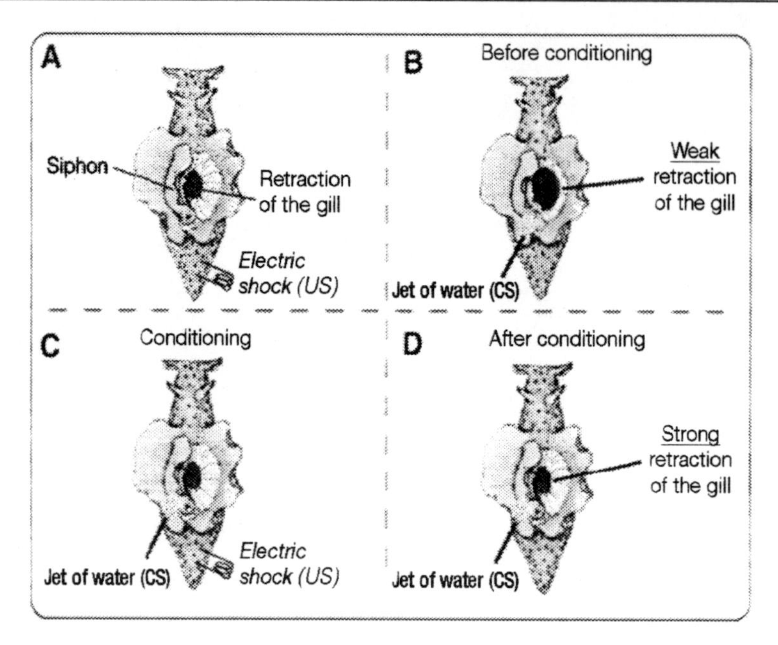

Figure 4.1 Conditioning in Aplysia californica (adapted from Kandel 2001)

of learning shared with more evolved species, including man. Specifically, it is possible to demonstrate in Aplysia a form of associative learning.

As its name indicates, what we mean by *associative learning* is the individual's ability to constitute associations among events. For example, we have all learned that the green light is the one at which we get moving, and that we have to stop when the light turns red. In this case we associate a particular visual stimulus, a color, with a voluntary motor response. But associative learning involves more sophisticated forms, especially classical conditioning and instrumental conditioning.

In the case of classical conditioning described by the Russian physiologist Ivan Pavlov (1927), the subject associates a stimulus that triggers a measurable response physiologically—in the case of Pavlov's dog, the presentation of food that triggers salivation—with another stimulus, neutral in itself, for example the sound of a bell. If, in the course of conditioning sessions, the sound of the bell is systematically associated with the presentation of food, the dog will salivate as soon as the bell is rung even in the absence of food. The animal has learned to associate an unconditional stimulus (US), that is, a stimulus (the presentation of food) that triggers a response (salivation) without conditioning, with a conditional stimulus (CS), that is, a stimulus (the sound of a bell) that triggers a response (salivation) after conditioning, hence an act of learning.

The sequence of presentation of the US and the CS is decisive for this kind of learning: for conditioning to take place, CS must always come several seconds before US. In other words, the sound of the bell must faithfully predict the presentation of food. If CS follows US or precedes it by too long an interval of time, for example thirty seconds, the conditioning will not occur. There must somehow be temporal coincidence between the two stimuli for conditioning to be established. This notion of temporal coincidence is essential, and we shall see that it has a counterpart both on the level of cellular and molecular mechanisms of plasticity and, surprisingly, in the Freudian model of the experience of satisfaction.[1]

1. For Freud (1895, 1926), an infant left alone, without recourse, is overcome by a state of distress. For example, the sensation of

Let us return to Aplysia. One form of classical conditioning in Aplysia involves an electric shock to its tail area (the unconditional stimulus, US; see Figure 4.1) and the reflex retraction of the gill (response); the conditional stimulus (CS) is represented by a discrete jet of water on the mollusk's siphon, located in its dorsal region. In the absence of conditioning, this jet of water, like any CS, triggers only a very weak reflex response of retraction of the branchia (Figure 4.1.B). In contrast, after several sessions of conditioning, in the course of which the US (electric shock) is preceded by the CS (jet of water) according to the Pavlovian protocol (Figure 4.1.C), Aplysia responds by a massive retraction of the

---

hunger it experiences leads to increasing internal tension associated with unpleasure. The resulting state of distress cannot be calmed by an action on the part of the infant alone, for the child does not yet have the sensorimotor skills to feed itself. Its only possible action is to cry. The other, in this case its mother, responds to its crying with her breast, calming and feeding. In this way an association is established between crying and being calmed by the other. For this association to leave a trace, it is essential that the crying and the specific action of the *Nebenmensch* coincide temporally in a *Gleichzeitigkeit* (simultaneity), as Freud (1895) strongly emphasizes. This is reminiscent of Pavlovian conditioning, even if the conditional stimulus (the sound of the bell) comes from the external world, whereas, for the infant, it is internal reality, the sensation of hunger, that triggers the crying. This simultaneous coinciding of two stimuli, one internal (distress) the other external (the specific action of the other leading to food), leaves a trace, an inscription, that undoubtedly has a synaptic basis.

branchia to the sole application of the jet of water to the siphon (Figure 4.1.D).

What, then, are the neuronal mechanisms that made it possible to establish this association, in other words, this learning? In order to understand them, let us study the synapses transmitting information in this simple neuronal circuit.

First of all, a sensory neuron is stimulated by the jet of water (CS) to the siphon area. It makes a synapse with a motor neuron that commands the retraction of the branchial muscle. The mechanisms of information transfer (see chapter 2) are of course at work in this circuit,[2] but a third neuron, which we shall call a modulator, is also in play: its axon makes a synapse on the presynaptic terminal of the sensory neuron and releases a neurotransmitter, serotonin, when the unconditional stimulus (US) is delivered (electric shock) (Castellucci and Kandel 1974).

The serotonin receptors are of the metabotropic type; their activation causes the formation of a second messenger, cyclic adenosine monophosphate (AMP), at the presynaptic terminal, that, like all second messengers, can

2. Thus the stimulation of the siphon triggers action potentials along the axon of the sensory neuron, releasing at its terminal, in a calcium-dependent manner, a neurotransmitter that, in turn, triggers excitatory postsynaptic potentials in the motor neuron. The depolarization of the motor neuron generates action potentials in its axon, which, invading the terminal, cause the release of the neurotransmitter. The effect of the neurotransmitter on the muscle, in the case of the CS, is to trigger the very slight contraction.

modify the activity of ionic channels. At the presynaptic terminal of the sensory neuron, it is the activity of a potassium channel that is modified, the effect of which is to depolarize the terminal. If the conditional stimulus (the jet of water on the siphon) precedes by several thousandths of a second the US (electric shock), the effect of the serotonin released by the modulatory neuron will manifest itself in a terminal invaded by an action potential in which, as a result, the concentration of calcium is increased. Now it happens that in the presence of calcium the activation of the serotonin receptors produces much more cyclic AMP, which results in depolarization of the presynaptic terminal in a more important way and, in consequence, in the release of more neurotransmitter, which then, via the motor neuron, leads to a significant muscular contraction.

In sum, the association of the two events (electric shock and discrete stimulation of the siphon), following a precise protocol (the CS must precede the US by a maximum of five hundred thousandths of a second), creates at the cellular level the conditions for a temporal coincidence between the increase in cyclic AMP and calcium in the presynaptic terminal of the sensory neuron. This coincidence increases the release of the neurotransmitter, thereby making the synaptic transmission more efficient. After several sessions of conditioning, this change in synaptic efficiency becomes permanent, and the animal massively contracts the branchial muscle in response to the CS, the stimulation of the siphon by a jet of water that before the conditioning protocol had very little effect. The animal has learned a simple behavior, and a form of memory has been constituted (Carew and Sahley 1986).

Let us now look at a form of synaptic plasticity in a neuronal circuit involved in behaviors even closer to what can be observed in man. After all, stimulation of the siphon and retraction of the muscles of the gill are not really part of the human repertoire! A type of behavioral experiment in the laboratory rat helps us here, allowing us to make a connection with what we have seen by way of associative learning in Aplysia.

The experimental setup in question is actually very similar: a sound of a given frequency is presented just before the administration of a weak, but unpleasant, electric shock to the animal. After several sessions the rat associates the two events and implements an avoidance strategy at the presentation of the sound in the absence of any electric shock. The rat has learned that the sound predicts the emergence of a disagreeable sensation. This association is very specific, for the rat does not respond in this manner if a sound of a different frequency is presented (Weinberger 2004).

With the aid of microelectrodes implanted in the regions of the brain involved in the treatment of sensory information, the activity of the neurons during the learning phase is recorded, especially in the region of the hippocampus, which has been demonstrated (for example, in lesion experiments) to be essential for the processes of memory. At the beginning of the conditioning sessions the neurons of the hippocampus do not respond to the sound. In contrast, once the animal has learned the avoidance behavior merely at the presentation of the sound, the electrodes register sustained activity, in bursts, each time the sound is presented, even in the absence of any electric shock. The rat has learned to associate the sound to the electric shock, and we find a trace of this learning in the synaptic activity of the neurons of the hippocampus, in the form of a new activity absent before the conditioning.

Even more striking is that this synaptic "memory" is lasting (like the behavioral memory, that is, the avoidance strategy at the presentation of the sound): the presentation of the sound alone continues to trigger bursts of neuronal activity long after the learning sessions.

To explain the establishment of this synaptic trace left by learning, let us once again change the experimental protocol. Instead of presenting the animal with external stimuli (for example, a sound) and recording neuronal activity, another electrode is used to stimulate the neuronal circuits of the hippocampus directly. With this experimental setup it is possible to "condition" the circuit by applying stimulation of a high frequency, for example a hundred stimuli per second. As a result of this conditioning we observe that the target neurons respond much more intensely to subsequent stimulations mimicking the spontaneous activity of this neuronal circuit. In other words, the transfer of information at the level of the synapses has been facilitated by direct stimulation of the hippocampus. Synaptic transmission has been potentiated. This potentiation lasts for several weeks, indeed several months (hence its name, long-term potentialization, LTP (fig. 4.2.) (Bliss and Collingridge 1993; Bliss, Collingridge, and Morris 2003).

These initial experiments done on the animal (the rat) have thus demonstrated that by a simple stimulation, though to be sure one with special characteristics (brief and at high frequency), a neuronal circuit can be "conditioned" and the efficiency of synaptic transmission there lastingly increased (LTP). Nevertheless, the rat is

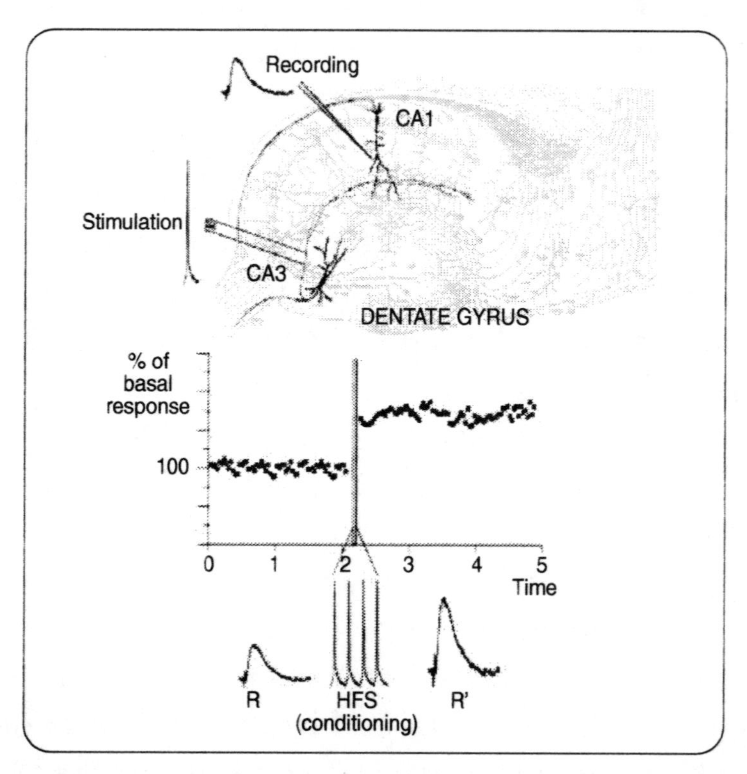

Figure 4.2 Model of synaptic plasticity: long-term potentiation (LTP) in the hippocampus. Following a high frequency stimulation (HFS), the response of the postsynaptic neuron is greatly increased.

not the Aplysia, which has large neurons, and it is difficult to make intracellular recordings in the rat as one can do in the case of the Aplysia. On the other hand, the hippocampus of the rat offers a particular experimental advantage: given its simple structure, slices of this region can be prepared, preserving the organization of the synaptic circuits (fig. 4.2). These sections can be kept alive for several hours while the neuronal circuits are stimulated and the synaptic activity (including the modifications of its efficiency) is studied.

In particular, it is possible to stimulate presynaptic axons, triggering action potentials, and record the responses of the postsynaptic neurons (Bliss and Collingridge 1993; fig. 4.2). First the amplitude of these responses after test stimulation is noted, then the axons are "conditioned" by a high-frequency stimulation (around fifty stimulations at a frequency of one hundred per second), and the responses of the target neurons are recorded once again. What is noted is a significant increase in the postsynaptic responses and the establishment of a long-term potentialization (LTP). The synaptic efficiency of this hippocampal circuit has been lastingly modified, and a synaptic trace is established.

Yet the modifications of synaptic efficiency constituting the LTP are observed in one and the same neuron only at the synapses established by afferents conditioned by high-frequency stimulation. In other words, the LTP is demonstrated solely at the synapses established by the axon terminals of afferents conditioned by high-frequency stimulation. Thus there is no global modification of the excitability of the postsynaptic neuron; were this the case, all the synapses received by a neuron would be facilitated by a stimulus conditioning one or the other of its afferents. What we are dealing with, then, is a phenomenon spatially limited to conditioned synapses.

More precisely, what happens when an intense stimulus is applied to a bundle of axons (which is done when a circuit of the hippocampus is stimulated at high frequency in order to induce the LTP)? First, each axon produces several action potentials that, by causing the neurotransmitter to be released from the presynaptic terminal, generate an EPSP at the postsynaptic neuron. The EPSPs generated by each action potential add up and produce a significant depolarization of the postsynaptic neuron. This is called the phenomenon of temporal summation (see fig. 2.2 in chapter 2).

What is more, the bundle stimulated to induce the LTP contains several axons. Hence the high-frequency stimulation activates several axon terminals simultaneously, thereby inducing several EPSPs. These EPSPs triggered at the same time add up. This phenomenon of synaptic integration is called spatial summation (see fig. 2.2). As with temporal summation, the result is a massive depolarization of the postsynaptic neuron. In other words, for a lasting modification of synaptic efficiency to be induced, it is necessary that several EPSPs (temporal summation) occur at several synapses (spatial summation) during a time window of several milliseconds, thus leading to a massive depolarization of the postsynaptic neuron. This is exactly what is produced by the high-frequency stimulation generating the LTP (Markram et al. 1997).

Let us move on to the physiological conditions leading to lasting increase in the synaptic efficiency connected with the establishment of a mnemic trace, returning to the case of our rat who has learned to associate a sound of a given frequency (stimulus 1) with an electric discharge (stimulus 2) that triggers an avoidance behavior. As with the other forms of associative learning, the experimental protocol requires that stimulus 1 precede stimulus 2. Each of the two stimuli activates specific hippocampal circuits. Let us hypothesize that certain axons of each of these two circuits converge on a postsynaptic neuron: the synapses they establish with it will be activated at the same time when the two stimuli are presented according to the protocol. This *twofold temporal coincidence of a spatial convergence* in fact generates a massive depolarization of the postsynaptic neuron, a much more significant one than if the axons activated by a single stimulus (for example, stimulus 1) were active. The association of two stimuli within a limited temporal window, indispensable in any form of conditioning or associative learning, thus establishes—physiologically this time—the conditions for the production of a lasting potentialization of synaptic transmission. According to Donald Hebb (1949), the Canadian psychologist who first hypothesized the fundamental concepts of synaptic plasticity as the cellular basis of the mechanisms of memory, the neurons active at the same time are those that establish associations among themselves. As he puts it, "neurons that fire together wire together."

In sum, the neuron on which converge signals (action potentials) generated in circuits activated by stimuli coming from associated events acts as a coincidence detector (Markham et al. 1997). By what molecular mechanism? Here, too, recent advances in the experimental neurosciences have offered an explanation, perhaps not the only one, involving the mode of functioning of receptors of glutamate, the principal neurotransmitter of the nervous system, which is released at the synapses of the hippocampus that are subject to synaptic plasticity of the LTP type. Several subtypes of glutamate receptors have been identified. Two of these are especially involved in synaptic plasticity: receptors of the Alpha-amino-3-hydroxy-5-methyl-4-isoxazolepropionic acid (AMPA) type and those of the *N*-methyl-D-aspartate (NMDA) type, these two acronyms designating the pharmacological compounds acting selectively on one or the other subtype of receptor, whereas glutamate, obviously, acts on both of them.

The AMPA and NMDA receptors are of the ionotropic type (see chapter 2), which is to say that the binding of the neurotransmitter to the receptor triggers the temporary opening (lasting several thousandths of a second) of a channel permitting the passage of ions across the membrane.

For the AMPA receptors it is sodium ions that pass, hence positive charges. This process will render the membrane potential of the postsynaptic neuron less negative and generate an EPSP. The postsynaptic neuron will thus be depolarized each time glutamate is bound to the AMPA receptors. This depolarization, reduced though it be, is the one observed when a neuronal circuit is activated by a physiological stimulus (the stimulus 1 of our example).

The NMDA receptors have several particular properties that distinguish them from the AMPA receptors. First, though these are ionotropic receptors, the binding of glutamate does not automatically trigger the opening of a transmembrane channel. Another condition must be met: the neuron on which the NMDA receptor is located must have been depolarized. For, in the absence of depolarization, the channel that is an integral part of the NMDA receptor is somehow "corked up" by magnesium ions. Depolarization "uncorks" the magnesium, as it were, making the channel permeable.

Here we have another difference between AMPA and NMDA receptors: the latter are permeable not only to sodium but to calcium as well. Now calcium, as we shall see, not only has positive charges that contribute to the amplification of the EPSP; it also functions as an intracellular signal that triggers the processes essential to the long-term establishment of synaptic plasticity.

Thus, if the NMDA receptors are to be active, two events must occur simultaneously: glutamate must be bound to the receptors, and the neuron on which these NMDA receptors are located must be depolarized (Markram et al. 1997). This occurs, for example, when two neuronal circuits converging on the same neuron are simultaneously active during a limited temporal window. This twofold temporal coincidence of a spatial convergence of activated axons generating a strong depolarization of the postsynaptic neuron will thus free the NMDA receptor of the blockage due to magnesium. Hence the NMDA receptor will be activated by the glutamate present in the synapse, causing a massive entry of calcium into the postsynaptic neuron. The NMDA receptor is thus a molecular operator sensitive to the simultaneous emergence of a massive postsynaptic depolarization and the activity of presynaptic elements. The NMDA receptor, therefore, serves as a coincidence detector: coincidence in the activity of converging circuits and coincidence among activities or pre- and postsynaptic elements.

The results of a massive increase in the concentration of calcium are many, some working short term (minutes), others long term (hours, days, months). In the short term, calcium activates an effector protein with the barbaric name of calcium-calmodulin kinase II, which triggers at least two mechanisms aiming at an increase in the sensitivity to glutamate of the postsynaptic neuron. The first of these is a phosphorylation of the AMPA receptors present on the postsynaptic neuron. Phosphorylation, the addition of a phosphate group to a protein, has the effect, in this case, of increasing the sensitivity of the AMPA receptor to glutamate. The second effect consists in promoting the insertion of new AMPA receptors into the membrane. Thus the synapses activated by the LTP protocol or by the simultaneous activation of convergent afferences become richer in AMPA receptors more sensitive to glutamate. In later stimulations the same quantity of glutamate will therefore produce a greater response, a larger EPSP, at these synapses, and the transfer of information will be facilitated.

But there are also mechanisms opposite to those that produce potentiation, namely those that lead to a decrease of synaptic efficiency, the experimental model here being long-term depression (LTD). For LTD, which also seems to involve calcium through intracellular increases clearly inferior to those observed for LTP, can be induced in the same hippocampal circuits as those subject to LTP by low-frequency stimulations (for example, five instead of one hundred stimulations per second). This stimulation protocol depolarizes the postsynaptic neuron in a more modest fashion. Here the magnesium block of the NMDA receptor is only partially removed, for not only does little calcium enter the postsynaptic neuron, but it does so with a different kinetics. To sum up, we may say that, with regard to the mode of entry of calcium into the postsynaptic neuron, the LTP produces a massive, oceanic wave, whereas the LTD produces the ripples on a lake. The reduced increase in calcium caused by the LTD protocol activates other enzymes, phosphatases, whose effect on the AMPA receptors is exactly the reverse of those mediated by calcium-calmodulin kinase II: reduction of sensitivity to glutamate and decrease in the number of AMPA receptors inserted in the membrane. The result is thus another form of neuronal plasticity shown by a decrease of synaptic efficiency (Bear 2003).

These, then, are the experimental data that substantiate the notion of neuronal plasticity, the fact that certain stimuli coming from the external world leave a trace in the neural network in the form of modification of synaptic efficacy. It seems entirely justified to speak of a trace, not only in terms of molecular mechanisms, but also in terms of a trace left by experience at the level of the very structure of the synapses. Subtle analyses, conducted with the aid of microscopic techniques enabling us to visualize the synaptic contacts in different experimental preparations, including those in living animals, have revealed such structural modifications. Recall that the contacts between an axon and a postsynaptic neuron basically occur at the dendritic spines. As a result of protocols inducing a synaptic facilitation, we see a structural modification of the contacts between axons and postsynaptic neurons, one that manifests itself as a duplication of the dendritic spines. The receptive area of the postsynaptic neuron is thus considerably enlarged (Lüscher et al. 2000; Bonhoeffer and Yuste 2002; see fig. 4.3).

This phenomenon, which occurs together with the other molecular mechanisms we have examined—increased sensitivity of the AMPA receptors accompanied by the insertion of new receptors into the membrane—contributes to the increase in efficiency of the transfer of information to the facilitated synapses.

If experience really leaves a trace in the neural network, a fundamental question remains: If these functional and structural modifications are indeed related to the establishment of mnemic traces, and hence of memories

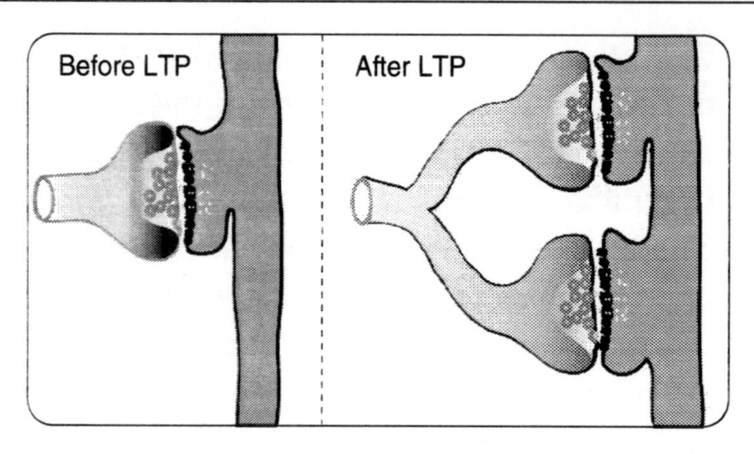

Figure 4.3 Duplication of the dendritic spines during synaptic plasticity. We may say that a trace is literally inscribed in the neural network by the structural modifications of the synapses.

and acts of learning persisting for several years, then we must postulate that they are, from the outset, permanent or at least durable by nature. But this is not the case. Specifically, the mechanisms of calcium-dependent phosphorylation, which play a role in the phenomena of plasticity, are as a rule rapidly reversible, in several minutes or at most several hours. In addition, the molecules of receptors and enzymes—AMPA or calcium-calmodulin kinase II, to mention only two—have a life span limited to several days or several weeks. Like all proteins, they are constantly being degraded and replaced. The mechanisms we have been describing up to now are in fact essential to *induce* the mechanisms of synaptic plasticity or to permit the establishment of modifications underlying short-term memory. But other mechanisms come into play for long-term memory, and this leads us to make a short (very short) detour into molecular biology, in particular into the mechanisms controlling the expression of our genes.

The human genome, which has been sequenced (Guttmacher and Collins 2003), comprises around thirty thousand genes, of which approximately half are specifically expressed in the brain (Insel and Collins 2003). *Expressed*: this is a key term for the understanding of the molecular mechanisms of synaptic plasticity persisting over the long term.

For the genetic material inscribed in the form of deoxynucleic acid (DNA) is entirely contained in the nuclei of all the cells in the human body. In other words, the sequences of the DNA of the thirty thousand genes of the human genome are present in each of our cells. The fact of being present does not necessarily imply that every gene is expressed in the form of protein coded in each cell. For example, the gene coding for the pigmented protein coloring the iris of our eyes is not expressed in the cells of the nervous system, although, like every other gene, it is present in the DNA contained in the nuclei of the neurons. For the genes are *expressed* selectively depending on the type of cell.

Moreover, the level at which a gene may be expressed by a cell varies according to the effect of regulatory elements, called promoters, present in the DNA sequence. Here again the rheostat is a useful image. When a promoter is active, the production of a protein by the gene coding for it will be stimulated. That is to say, the levels at which a cell expresses a protein are not fixed; they are subject to variations over the course of time. Thus there exists what might be called a plasticity in the expression of genes. If the promoters are the locks opening the doors of the expression of genes, the keys are little molecules, called transcription factors, that bind in highly specific fashion to the DNA sequences represented by the promoters.

One of these transcription factors has been especially studied for its role in the establishment of long-term plasticity: CREB (cyclic AMP responsive element binding protein). Let us unpack this barbaric name to find what we need to know about CREB. To start with, let us look at *cyclic AMP*. As we saw in chapter 2, this is a secondary messenger, that is, a molecule formed inside a cell as a result of an extracellular signal, for example a neurotransmitter (the primary messenger) interacting with a membrane receptor. Continuing with *responsive element binding protein*: where is this responsive element? On the DNA. A protein binding to DNA? Does this remind you of something? Yes, of course; this is the very definition of the transcription factor. CREB is thus nothing other than a transcription factor. When it is activated by cyclic AMP, it binds onto a DNA sequence called CRE (cyclic AMP responsive element), thereby activating the expression of a given gene and hence of a given protein. Let us recall that, via the mechanism of transcription, RNA (ribonucleic acid) is produced from DNA. And then follows the stage of translation, which leads to protein synthesis from RNA.

But how, you will ask, are these subtle molecular mechanisms leading to the control of protein synthesis involved in long-term memory? Let us go back forty or so years to the time of the Beatles or Elvis Presley. "Yellow Submarine," "Love Me Tender": these songs have stood up to time. For those of us who were teenagers in the 1960s, they are well and truly anchored in long-term memory systems. And it is from this period that the experiments are dated showing that the intracerebral injection of substances capable of inhibiting protein synthesis considerably reduced the consolidation of long-term memory. These were crude experiments, to be sure, but they indirectly showed the role of the regulation of gene expression in the processes involved in long-term memory. In a similar way, the long-term maintenance of LTP is strongly inhibited when the mechanisms of transcription and translation are blocked.

We can say, then, that synaptic plasticity depends on the modification or mobilization of the existing proteins (for example, the phosphorylation of channels or the mobilization of receptors) for its induction and short-term establishment. In contrast, the synthesis of new proteins through subtle mechanisms controlling the expression of genes seems to be essential for its long-term consolidation (Lamprecht and LeDoux 2004).

But are synaptic contacts the only elements involved in neuronal plasticity? Are the dendritic spines the only ones to be duplicated? After all, it might have been thought that new neurons would be produced in the course of learning; communication between neurons would have been potentiated in this way. Such a claim would have run counter to established dogma until seven or eight years ago. Wasn't it said that each individual is born with a given reserve of neurons that simply diminishes over the course of life? A sad prospect. But today we know that neurogenesis occurs. Thus the adult brain constantly produces new neurons from stem cells. These are largely undifferentiated cells that, as a result of various factors, can become differentiated into neurons (Kempermann, Wiscott, and Gage 2004). One of these factors is learning (Shors et al. 2001). Even physical exercise stimulates neurogenesis (Van Praag et al. 1999), the mechanisms of which have been especially well studied in the hippocampus. The question that remains nowadays is how the newly generated neurons connect up with the neurons already in place and what role they play in the establishment of new mnemic traces.

The mechanisms of plasticity associated with experience are not the sole prerogative of neurons. For other types of cells are present in the brain, in particular the glial cells, which are in fact five to ten times more numerous than the neurons. A special type of glial cells, the astrocytes, surround practically all the synapses with their processes. The astrocytes capture glucose from the vessels irrigating the brain, providing the neurons with fuel on demand (Magistretti et al. 1999). It has been shown that, during the fine motor learning that especially engages the neuronal circuits of the cerebellum, not only does the number of synaptic contacts increase but the astrocytes, too, participate in the mechanisms of plasticity by a considerable augmentation of the surface of the processes surrounding the synapses (Jones and Greenough 2002).

Ultimately, we see that the trace left by experience is associated with structural and functional modifications of the synapses and their efficiency, whose subtlest cellular and molecular mechanisms we are beginning to know today. Other mechanisms, such as neurogenesis and the role of glial cells, are slowly coming to light.

5

Forgetting the Name Signorelli: *Synaptic Trace and Psychic Trace*

It all seems simple: experience leaves a trace in the synaptic network. The mechanisms responsible for this synaptic trace are those of plasticity. To put it clearly, the transfer of pieces of information between neurons works more efficiently at facilitated synapses, either because a greater amount of neurotransmitters (glutamate) is released from the presynaptic terminals or because the mechanisms underlying the postsynaptic responses are more efficient, making the postsynaptic neurons more reactive. As we have seen, the mechanisms of postsynaptic facilitation are limited to one or several synapses at each neuron. This is not, in fact, the response of the ten thousand synapses found on a neuron potentiated by LTP. Only the few synapses where convergence and simultaneity of activation have engaged the mechanisms of plasticity and facilitation are affected.

But how do we move from several dozen, or several hundred, indeed thousands of facilitated synapses in the course of an experience to the representation of this experience? To go back to our example of Christmas in chapter 2, how does the representation of the Christmas tree get inscribed in the facilitated synaptic network? The answer to this question is far from obvious. The results allowing us to sketch out an answer are still fragmentary, and their integration into operational models are basically hypothetical in nature. But nevertheless let us try, perhaps more by way of analogy than as pure fact, to describe the models that could explain the neural bases of representations.

Each of the hundred billion neurons constituting our brain is connected to other neurons across approximately ten thousand synapses (Bear, Connors, and Paradiso 2001). One hundred billion neurons times ten thousand synapses make one million billion synapses (10^{15}). That's a lot of synapses! Let us imagine that the facilitation of a thousand synapses occurred during the association between an object and an event, for example a fir tree and Christmas festivities. This association is repeated regularly, every year at the end of December, and is thereby consolidated. For we know that the repetition of an association, or even the mere bringing back into awareness of this association, consolidates the memory (Squire, Stark, and Clark 2004). We therefore posit that the facilitation of 10^3 of the 10^{15} synapses would correspond to the representation of an object/event, an experience which is that of the Christmas tree. If we could number each of the 10^{15} synapses, we could posit that synapses 15; 27; 145; 1,890; 100,238; and so forth are facilitated, until we arrived at a total of one thousand synapses out of the set of 10^{15} so as to code the image of a Christmas tree. Although this hypothesis still does not explain how the image is formed or visualized, it provides a model for coding an object/experience in a unique pattern of facilitations of a certain number of synapses. For another object/experience a different unique pattern of synaptic facilitations could serve as a neural substrate for the representation. The possibilities are practically infinite, as are our experiences.

Certain neurobiological theories regarding the neural substrates of representation propose the existence of such sets of neurons at which specific synaptic facilitations occur. Following Hebb's (1949) hypotheses on sets of neurons, many studies have suggested a structural model of neural representation corresponding to elements of external reality (McNaughton 2003). A recent formulation of this theory, based on observations of the visual system, has been put forth by Wolf Singer (1998, 2004; Edelman 1992). This view suggests that there are neuronal assemblies, dynamic associations among sets of neurons, defining a constellation of characteristics peculiar to a given object or experience.

In other words, these "metarepresentations" are constituted by the temporary dynamic association of neurons organized into functionally coherent sets. From this perspective, the synchronous activation of these sets for several milliseconds is related to specific representations of elements of external reality. From this notion of synchronization in the activity of these sets of neurons, it follows that attentional mechanisms would facilitate this synchronization and hence representation (Fuster 2000). In other words, the representation of external reality is in some sense mapped at facilitated and distributed synaptic networks that can be temporarily reactivated. Thus, to push this paradox a bit further, there would be not one representation, one memory, inscribed in a synapse but a network of facilitated synapses activating dynamically, and this synchronous activation would correspond to a representation of a specific experience of the external world (fig. 5.1).

Let us take another image: a skyscraper at night, the offices are empty, the facade is dark. Suddenly, one after another, the windows light up and small rectangles of light break into the darkness. The process seems random at first, but as the lights appear, a form takes shape and becomes clear. All of a sudden it's there: a Christmas tree, drawn on the facade of the skyscraper by hun-

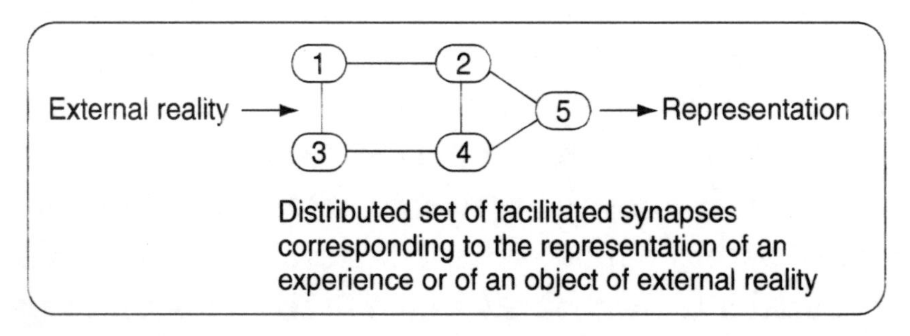

Figure 5.1 Distributed set of facilitated synapses corresponding to a metarepresentation

dreds of little lights. A precise coding in the pattern of the illumination defines an object representing a Christmas tree. The facilitated synapses of the brain do not, of course, "draw" a pine tree when the memory comes into awareness. On the contrary, it is likely that a pattern of precise, synchronous synaptic activity is activated when the memory of a Christmas tree materializes: there is a synaptic coding of the representation.

We can see how, via these synaptic mechanisms, a mental trace or representation of an object/event of an experience could be constructed. But can we generalize this process to the unconscious mental traces through which an unconscious internal reality would be constructed, such as the fantasy scenario, for example, whose exploration is at the center of the psychoanalytic enterprise?

Let us return to Freud's letter to Fliess of December 6, 1896, in which Freud begins to schematize the mental apparatus (1887–1902, 174; fig. 5.2). At one end of the diagram is perception at the other, consciousness.

Figure 5.2 Freud's diagram in his letter of December 6, 1896, in which he posits an initial inscription (trace), the sign of perception (WZ), and its fate in later transcriptions as carried by the various systems: unconscious (Ub), then preconscious (Vb) and conscious (Bw).

Between the two is a whole series of successive transcriptions in the form of mnemic traces that, for Freud, constitute the systems unconscious and preconscious that we may regard as systems of memory based on synaptic plasticity.

Thus we first find perception (W). The first registration, the first trace of this perception, is, for Freud, the sign of perception (Wz, *Wahrnehmungszeichen*), formed as a result of temporal coincidence (*Gleichzeitigkeit*) and leading to simultaneous associations.[1] We have seen (chapter 4) how the coincidence of stimuli is necessary for the establishment of a lasting modification of synaptic efficiency, that is, a synaptic trace. First there is the experience and its perception, then, according to

1. For Freud, the sign of perception is wholly unable to become conscious. Yet we may wonder why he makes this statement in this initial formulation. The question is all the more relevant because the model of the psychic apparatus he constructs in his second schematization (on the different models of the psychic apparatus see Lacan 1954–1955) includes a system perception-consciousness implying a direct inscription of perception, accessible to awareness, even if it may also be inscribed on the level of the unconscious at the same time. This latter outcome has now been confirmed by experimental data showing the treatment of the sensorial modalities by subcortical networks, especially the projections of the thalamus to the amygdala, which also implies a non-conscious treatment of sensory information (LeDoux 1996). The relations among perception, consciousness, determination, and the traces inscribed in systems not accessible to awareness are discussed in detail in chapter 13.

Freud, the sign of perception, the first psychic trace that it is possible to parallel with the synaptic trace (fig. 5.3).

It is striking how, through an approach that places psychoanalysis in tension with the general linguistics of Ferdinand de Saussure, Lacan constantly stressed the fact that the sign of perception must be given its real name: the signifier.[2] To continue in this line of reasoning, this signifier would correspond to a modification of synaptic efficacy (in certain specific synapses) in relation to a unique lived experience, which would be its *signified*. We could thus place on the same level the modification of synaptic

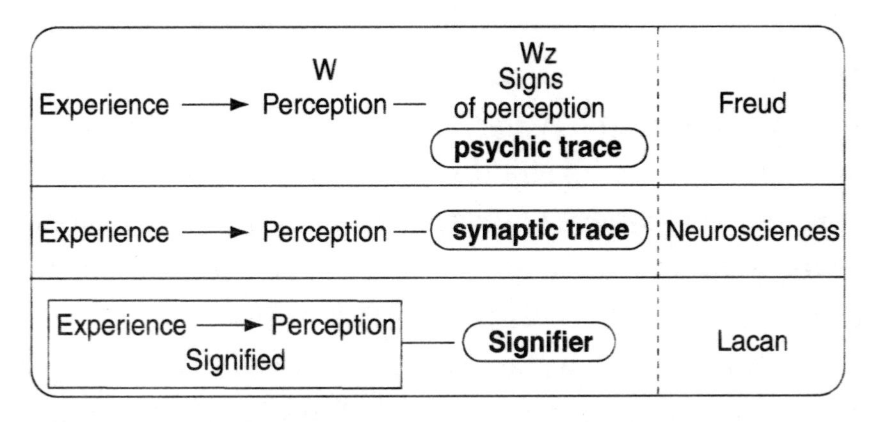

Figure 5.3 Convergence among the notions of psychic trace, synaptic trace, and signifier

2. "[W]e can immediately give to these *Wahrnehmungszeichen* their true name of *signifiers*" (1964, 46). This assimilation of the *Wahrnehmungszeichen* to the signifier is not an isolated case. By way of example we can cite the following passage: "But don't forget that we are dealing with the system of the *Wahrnehmungszeichen*, signs of perception, or, in other words, the first system of signifiers, the original synchrony of the signifying system" (Lacan 1959–1960, 65).

efficacy (the synaptic trace according to neurobiology) and the signifier (according to Lacan). These three terms—sign of perception, synaptic trace, and signifier—would correspond to a signified that is nothing other than the perception of the experience of external reality.

In linguistics, a signifier is a sequence of letters in relation to a signified, be this an object or an event (experience). In other words, a signifier, in the linguistic sense, is directly associated with a signified that is the experience itself. Now if this signified corresponds to the synaptic trace, and if the concept of synaptic trace sums up the model of a pattern of synaptic facilitation corresponding to the representation of an experience (recall the lights on the skyscraper), if we pursue this line of reasoning and follow to the letter (indeed!) the correspondence between signifier and synaptic trace, it is possible that the pattern of synaptic facilitation (the numbers of the activated synapses) is the linguistic equivalent of a sequence of letters constituting a signifier.

Thus experience and language would be linked at this fundamental juncture that is the sign of perception or synaptic trace.

Following this path, we see that the sign of perception materialized in the synaptic trace becomes the first fundamental juncture between language in its signifying articulation and living matter. And here we have everything that is at stake in what is peculiar to man as a linguistic being, that is, a being subject to the signifier beyond a simple language code.

Let us return to Freud's construction, focusing now on what becomes of the sign of perception. Freud believes that this sign can be reinscribed in other systems, thereby leading to later transcriptions in the unconscious (*Unbewusstsein*). If these reinscriptions occur through the mechanisms of synaptic plasticity and are organized in accordance with other associations—for Freud, "perhaps according to causal relations"[3]—they somehow constitute secondary traces that, in turn, will associate with each other to form new traces.

In this way the circuit of perception, memory, and awareness will be refueled several times, either directly from perception or from its reactivation in the successive transcriptions of the sign of perception. The experience perceived and then inscribed is thus transformed and deformed by a whole interplay of connections and associations, which leads to what might rather paradoxically be called an "endopsychic perception" in contrast to the perception of external reality (Lacan 1959–1960, 49), that is, an unconscious internal reality playing an afferent role, causing interference with the pole of awareness in the psychic apparatus (see chapter 9).

The combinations of inscriptions and retranscriptions can go on indefinitely. Let us take, for example, a series of words associated with the episode of inhibition Freud experienced on the shore of Lake Trasimene (see

3. Freud 1887–1902, 174. In chapter 14 we shall return to the issue of what this causality might be like.

chapter 3), such as *Trasimene, Hannibal, Rome, journey.* Each of these signifiers (each of these sequences of letters) is associated with a signified, a particular external reality proper to the individual. But the set of this sequence of signifiers can also evoke a new signified, for example a memory of childhood or, why not, a fantasy implied by this type of association. In other words, these signifiers, to each of which corresponds a synaptic trace (to the extent to which these terms have been committed to memory), are associated with signifieds in the reality of the external world, but at the same time each of these signifiers can be associated with other signifiers, produced by the same mechanisms—*cap, sidewalk, Jew*—that taken together contextually produce new signifieds, such as *father's humiliation, academic compromise.*[4]

Such a mechanism of association among signifiers, that is, association among traces (mental and synaptic), can participate in the organization of a fantasy in the unconscious, in this case one centering on the son's relation to the father, in accordance with various, highly invested experiences. Thus, as Freud suggests in Draft M, fantasies "arise from an unconscious combination of things experienced and heard, constructed for particular purposes" (1887–1902, 204). From a signifier that in reality is related to a precise signified, a new signified may be created in the unconscious in association with other signifiers.

4. This well illustrates the gap between external reality and unconscious internal reality.

There remains the question of inscription. On the conscious, cognitive level, the sequence of words and letters faithfully represents lived experience. On the other hand, this same sequence of words and letters can be associated on the unconscious level with other traces (other signifiers) and become organized into a chain of signifiers corresponding to a new signified, belonging to fantasy life, that no longer has anything to do with the event perceived in reality. At the same time, fantasy life constitutes a new influence on the organization of the neural network. In other words, a signifier, that is, the synaptic trace of an experience, can become associated with other signifiers that come from other experiences, resulting in a new signified that no longer has anything to do with the signifieds initially inscribed (fig. 5.4).

If the metarepresentation (fig. 5.1) of an experience corresponds, according to the theory of neuronal ensembles, to the synchronous activity of a whole set of synapses in which facilitation has been produced, a trace has thus been inscribed. We could therefore decompose the traces left by perceptions, each of which corresponds to a signified of external reality, into a series of inscribed and associated traces schematically representing neuronal ensembles in our diagram (fig. 5.5), as solid squares for trace 1, open squares for trace 2, open triangles for trace 3, and circles for trace 4.

The mechanisms of association form new neuronal ensembles that arise from the association of these primary traces and produce new signifiers, for example signifier A, constituted this time from the association of

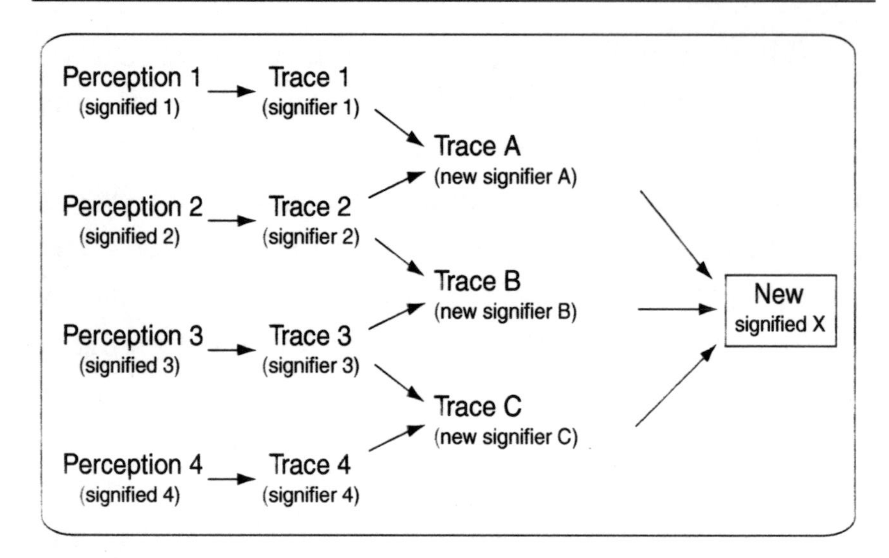

Figure 5.4 Each perception (1, 2, 3, 4, . . .) is inscribed in the form of a synaptic trace (or signifier 1, 2, 3, 4, . . .). In this first stage, each signifier corresponds to a signified in external reality. Through the mechanisms of synaptic plasticity, these primary signifiers can become associated with each other and reinscribed in the form of new traces (or signifiers A, B, C, . . .) that, via a signifying chain, produce a new signified X that may not correspond at all with the signifieds of external reality.

elements of trace 1 and trace 2, that are represented on the diagram by two solid squares and an open one. This process can be extended to other associations of traces into new signifiers. This signifying chain will lead to other associations represented by the neuronal ensembles constituting the new signified X.

As we see it, these new signifieds constitute elements of the fantasy scenario belonging to the unconscious internal reality of each person. This scenario can interact with the assessment of external reality, making it enig-

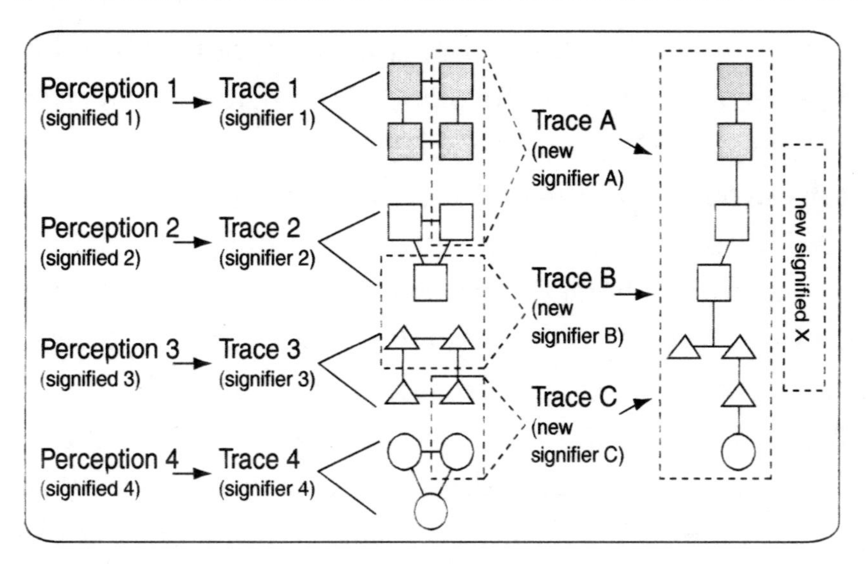

Figure 5.5 Schematic representation of associations and reassociations among certain elements of the initial traces. They constitute new ensembles corresponding to new signifiers whose combination constitutes a new signified X that no longer has a direct relation with the signifieds of external reality inscribed in the initial traces.

matic for the person and interfering with consciousness. Such short circuits are frequent in psychic life. A striking example is offered by Freud himself in connection with his forgetting the name *Signorelli* as recounted in *The Psychopathology of Everyday Life* (1901; see also Freud 1898).

Freud was traveling by train from Ragusa in Dalmatia to a station in Herzegovina. During the trip the conversation with the man seated in the same compartment turned to Italy. Freud asked this man if he has already been to Orvieto and seen the famous frescoes of . . . . Freud could not recall the name. Two names of painters came to mind, Botticelli and Boltraffio, but he couldn't manage to remember the name of the painter of the frescoes, Signorelli. Freud associated this forgetfulness with a disturbance produced by the preceding conversation. He had been discussing with his traveling companion the customs of the Turks living in Bosnia and Herzegovina. A colleague had told him that these people were full of confidence in medicine and fully resigned to death. When told that a relative was at death's door, they reply, "*Herr* [Sir], what is there to be said? If he could be saved, I know you would have saved him." Hearing this story, Freud had intended to tell another anecdote he recalled. These same Turks of Bosnia and Herzegovina attach exceptional value to sexual pleasure. When stricken with sexual problems they are seized with a despair that strongly contrasts with their resignation in the face of death. One of these patients said one day, "*Herr*, you must know that if *that* comes to an end then life is of no value" (1901, 3; emphasis in original).

Yet Freud held back from broaching this subject with his traveling companion. An idea occurred to him at the same time by way of association, an idea he did not want to think about. He remembered that, several weeks earlier, during a brief stay in Trafoi, he had received the news that one of his patients, who had given him a lot of trouble, committed suicide because he suffered from an incurable sexual complaint. It was under the influence of this reminiscence that Freud connected the name of the city he was in, Trafoi, with Boltraffio.

Seen in this light, the forgetting of the name *Signorelli* no longer seemed accidental at all. Freud observed the effect of psychic motives here. He recognized that he wanted to forget something other than the name of the painter. He wanted to forget something concerning death and sexuality: the suicide of his patient suffering from impotence.

The substitution of the name *Signorelli* occurred thanks to a displacement along the combination of other names, all this without regard to the meaning or the acoustic delimitation of the syllables. The associative series—*Signor, Herr, Herzegovina, Bosnia, Botticelli, Boltraffio,* and finally *Trafoi*—finds meaning in another scenario, an unconscious one. This other scenario does not use the word as such but only fragments, indeed several letters, for example the "Bo" of "Bosnia," which is found in the "Bo" of "Boltraffio" and is associated with the town of Trafoi (fig. 5.6).

The function of the signifier in this example prevails over that of the signified. It is the death of his patient that Freud wants to forget, but what he actually forgets is the name of the painter. Forgetfulness happens through something other than what one wishes to forget. The signifier, indeed the letter, thus brings to light the issue of meaning when we admit the existence of this other theater that is the unconscious.

Hence when the brain perceives and inscribes in the form of a trace stimuli coming from the external world, leading to the construction of a psychic trace,

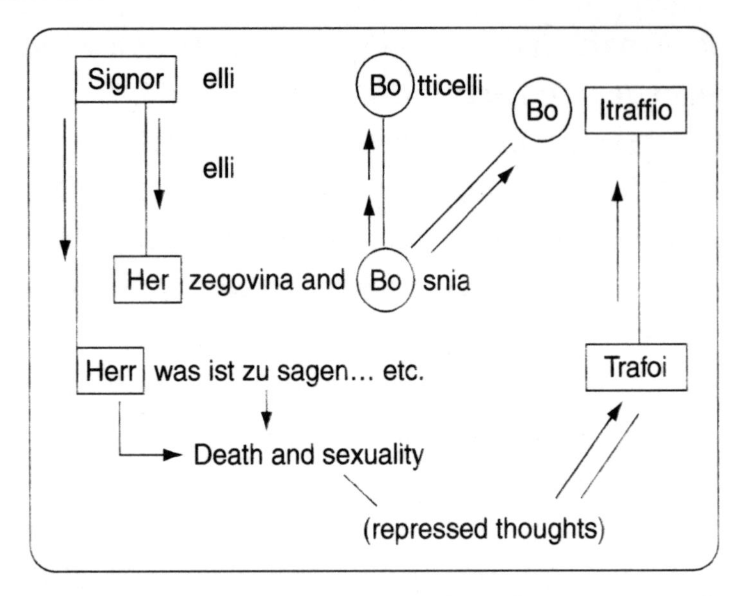

Figure 5.6 Diagram suggested by Freud to account for the mechanism of associations leading to the forgetting of the name Signorelli.

the transcription of an external reality, there may be a correspondence between the trace (signifier) and external reality (signified), the signifier corresponding to the signified, and this correspondence, which is conscious in nature and had to do with cognitive processes, constitutes the basis on which we anchor ourselves at several points in reality. In contrast, through successive transcriptions, the initial inscription may also leave the domain of consciousness and become a constitutive element of mental reality in the form of an unconscious representation.

The first part of the process corresponds to the inscription of external perceptions in the neural circuits by the mechanisms of plasticity. The second part involves a new transcription inscribed without direct relation to external reality and constituting a created element of a mental reality distinct from external reality.[5] The first phase is, in principle, conscious; it is the basis of learning and the construction of conscious, evocable memories. The second contributes to the formation of an unconscious inner reality underlying fantasy constructions. To return to the example of Freud, the pressure of what he did not want to know about the death of his patient led to a series of associations distancing him from it even as a connection was maintained through a concatenation of signifiers.[6]

5. External reality can thus become inaccessible as such, which connects with what Freud (1938) says about reality never being knowable.

6. The analysis of dreams necessarily involves taking into account these chains of signifiers that have no conventional relation to signifieds but that nevertheless make it possible, sometimes, to return from the mental reality to the experience lived in correspondence with reality via the construction and deconstruction of an unconscious internal reality.

6

Claire and the Pope: Perceptions and Emotions

Let us imagine that you recall a landscape dear to you, one whose evocation puts you in a state of great serenity, for example the image of those beautiful, fertile Tuscan hills where cypresses, olive trees, and vineyards weave a supple green tapestry to the horizon. With this memory and the impressions it brings firmly to mind, we are going to try to relate the cerebral processes mentioned for the constitution of representations—whether they can be recalled to consciousness or are unconscious in the form of fantasies—with somatic states involved in the emotion linked to these representations. What does the body have to do with all this?, you ask. Much more than one might think.

At each instant our brain has various sources of information. First there is the perception of external reality activating the sensory systems (touch, sight, hearing, etc.) and hence conscious information. But this external stimulation can also activate the unconscious internal reality formed through the mechanisms of plasticity peculiar to each person, beyond the external reality the person has experienced. This unconscious internal reality, which may be organized into fantasy scenarios, rearranges in a different way the representations conserved from perception without direct relation to the stimuli of external reality. Thus these representations may also be recalled to awareness by a stimulus coming from the external world, experienced "live." Finally, they may be reactivated by a voluntary or involuntary process with no relevant external stimulation.

Let us imagine that you are going into a room at a friend's house. You look at the furniture, the knickknacks, the pictures. You smell the roses placed in a large vase at the center of the table. You hear the notes of a piano someone is playing in the next room. Time stands still: you have an instantaneous image of the situation concretized thanks to your various sensory systems. But now, reactivated and recalled to your awareness, is a whole series of images stored in your memory systems. The pictures in the room remind you of an exhibit you attended recently. The roses recall a romantic tryst in your youth.

They remind you of Claire, how beautiful she was, how much she loved roses. The next link in the chain is the Piazza di Spagna in Rome, where you spent a lovely weekend with Claire. Rome, Michelangelo, the Vatican, the pope . . . . The pope is very ill; he has Parkinson's disease. This brings to mind the latest request for funding that you have to submit between now and Wednesday, having to do with a new hypothesis about the mechanisms of neurodegenerative diseases like Parkinson's. . . . The images are coming faster and faster to your mind when, suddenly, your friend's wife comes into the room. "Hi, how are you?" You return to the present; at this moment your primary sensory systems are active again: vision, hearing, smell. They are again in direct contact with the world around you.

Our life is somehow a permanent shuttling back and forth between the moment (when the primary sensory systems are in action) and the recall of representations (when the memory systems are active). You can also voluntarily trigger the mechanisms activating the sequence of representations in the absence of an external stimulus. This is what we did when we wrote the preceding lines.

But there is not only recall of representations. A phenomenon of another order accompanies perception and the recall to awareness of representations, namely the emotions that come into play, the feelings saved up along with their representation in the form of what Antonio Damasio (1994) calls somatic markers: a body memory of some sort. When you evoke the image of your former girlfriend, Claire, and the delicious weekend in Rome, a very pleasant feeling is associated with these images; this feeling could be defined as an emotion, the emotion of love. In contrast, when the image of the deadline for submitting the request for funding comes up again, you are now filled with a feeling of anxiety, a tension as unpleasant as it could be.

When you experience either the pleasant feeling associated with the image of Claire or the unpleasant one associated with the request for funding, you will note, if you are attentive, that the represented image is also associated with feelings more or less perceptible on the level of your body. The image of Claire will be able to evoke a strong feeling in the belly, the belly tied up in emotion, or even in the genitals. If you measured your blood pressure or heart rate, those might be increased. Likewise, the feeling of anxiety linked to the evocation of the upcoming deadline for submitting the request for funding is accompanied by somatic manifestations: increased heart rate, for example, or a fine sweat on the skin.

We are not often aware of the somatic states associated with the evocation of a representation or with a perception. Yet one could bet that, if Claire were suddenly to appear in the room, the sight of her would produce somatic responses even stronger than those activated by the representation of her image. The Scottish philosopher David Hume stated that the images we bring to awareness through memory are weaker than those directly produced by perception (Damasio 1994). The same could be said of the associated somatic responses.

Thus there are somatic states associated with a perception or a representation. Already at the end of the nineteenth century the American psychologist William James set forth a position that was very provocative at that time. A stimulus coming from the external world, though it activates the sensory system in question (sight, for example), does not just trigger a perception, according to James, but is also associated with a somatic response (change in heart rate, for example), and the simultaneous occurrence of an external stimulus and an associated somatic state is precisely the basis for the perception of an emotion. James (1890, discussed in Damasio 1994) gives several very telling examples. What would the feeling of fear be like, he asks, if we did not feel our heart beating faster, if we were not short of breath, if our lips were not quivering, and the like? We cannot, James says, imagine anger without a feeling of seething in our chests, or reddening of our face, or dilation of our nostrils, and so forth.

Now let us ask how a perception, or the evocation of a representation, comes to be associated with a given somatic state. The brain possesses a series of neural circuits we can group under the heading of *transductors* of a perception (or of the evocation of an image) into an emotion. One cerebral region plays a special role here: this is the amygdala (Aggleton 2000), a structure located on the inner side of the temporal lobe (the lateral part of the brain; see fig. 6.1) and receiving afferents from the primary sensory systems (vision, hearing, smell). In other words, a noise, the sight of an object, or an odor is able to activate certain neurons of the amygdala, in particular those located in the basolateral part. The neurons of this subregion of the amygdala are connected to, among others, certain neurons of another subregion of the amygdala, the central nucleus, and these neurons project massively toward the regions of the brain controlling the neurovegetative system.

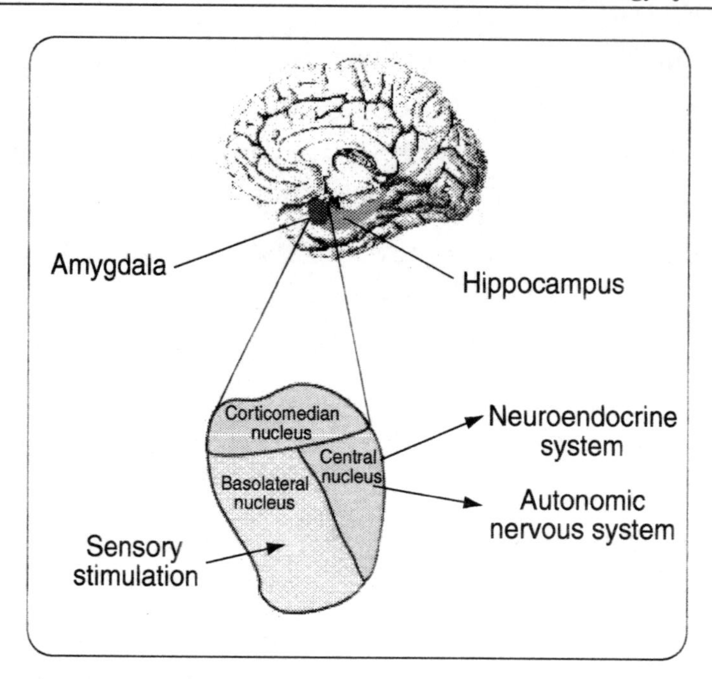

Figure 6.1 Neural circuits of the amygdala.
The amygdala receives sensory information and connects it with the neuroendocrine and autonomic nervous systems determining a given somatic state.

Now this neurovegetative system (parasympathetic and sympathetic) is itself composed of certain neural circuits that very efficiently control our internal organs and our hormonal system. In other words, heart rate, blood pressure, perspiration, the release of all kinds of hormones (for example insulin, which, in turn, controls the levels of glucose in the blood), gastric secretion, and intestinal motility—in short, all the mechanisms enabling an organism to maintain a physiological state, a state of homeostasis, are controlled by the neurovegetative system.

The English term for the neurovegetative system is very telling: autonomic nervous system. The term *autonomic* makes very clear that this system is not under voluntary control (unless one is a yogi who can control his intestinal motility and heart rate). In fact this system is not all that autonomous. We could say instead that it is *automatic* insofar as it is activated by stimuli from the external world not under voluntary control, stimuli translated by the neural circuits of the amygdala into somatic responses under the control of the neurovegetative system.

This is a rapid overview of some of the mechanisms that make possible the association of an external stimulus with a somatic state. The same mechanisms of transduction in the amygdala apply to representations recalled to awareness. Thus, when you were thinking of Claire just now, and somatic responses were triggered by this happy thought, it was the amygdala that played the role of transductor between the represented image and the somatic state. Here it would seem that the prefrontal cortex acts as an activator of the amygdala insofar as this cerebral region is involved in the temporary constitution of representations (see chapter 13). This prefrontal cortex (especially the medial and ventral parts), through projections to the amygdala, namely the basolateral nucleus (McDonald 1998), will activate the same neural circuits causing the neurovegetative system to come into play.

Yet the neurovegetative system is not the only effector modifying the somatic state. Another modality controlling

> the internal organs and glands of our body is the endocrine system, for this system, too, contributes to the homeostasis of the organism. In this case the amygdala has another role as transductor: another nucleus of the amygdala, the corticomedial nucleus, projects to the hypothalamus, the central control station of the endocrine system. Through the hypophysis, the hypothalamus controls the secretion of hormones acting on the various organs of our body. Hence, whether by a neuronal path via the neurovegetative system or by an endocrine path via the hypothalamus and the hypophysis, the amygdala influences the somatic state related to a perception or the recall of a representation (Brodal 1992).

But the brain needs to "read" the somatic state in which it finds itself as a result of a perception or the evocation of a representation. Working largely on the basis of subtle clinical observations, Antonio Damasio (2003) has identified some cerebral regions that at each moment detect the state of our organism. Of primary importance here are certain parietal regions of the sensory cortex, especially the insular region. At each moment ("live," as it were) these sensory regions in effect photograph the somatic state. The term *interoceptive pathways* has been used to describe these neural circuits that transmit bits of information from the interior of our body, in contrast to *exteroceptive pathways* like the visual, auditory, or olfactory systems that enable us to detect the state of the world around us. From the insula (see fig. 13.1 in chapter 13) the circuits of interoception project to, among other regions, the prefrontal, medial, and lateral regions of the cortex.

We have come full circle, then, and can describe a circuit leading to the association of a perception of the external world (or its evocation) and the perception of an associated somatic state. Let us review. The amygdala is activated from the primary sensory circuits; it modifies the somatic state through the neurovegetative and neuro-endocrine systems; the new somatic state is detected by the insula and the regions of the brain sensitive to circulating hormones: the association can be made between the external perception (or a represented image) and a somatic state (fig. 6.2; see Craig 2002).

We can return to the linguistic analogy made earlier, considering that a given somatic state can be associated with the signifier of a perception of external reality or a represented image. Or, like Damasio (1994), we can

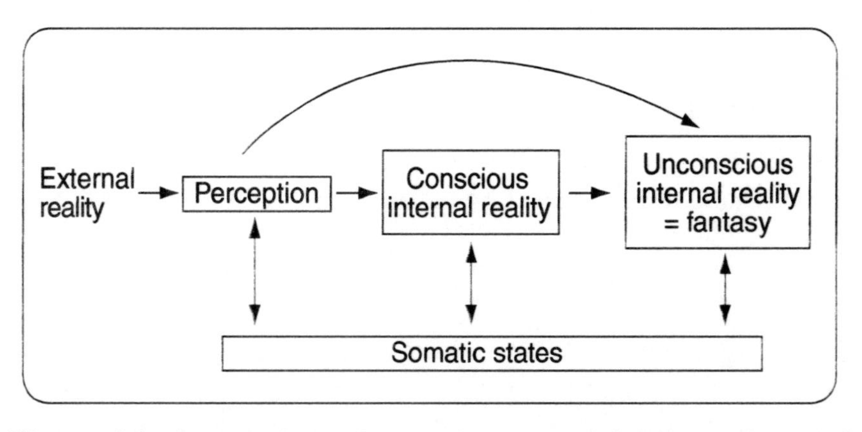

Figure 6.2 Association of somatic states with perception and its various conscious and unconscious inscriptions

use the term *qualificative* of the bodily state juxtaposed with a perception to say that this state is associated with a *qualified,* for example the face that triggers a certain emotion.

We might also posit that this somatic state can be associated with a trace of the experience, marking its fate in the defiles of unconscious internal reality as it becomes an integral part of the fantasy process (fig. 6.2).

7

Milk and the Sound of the Door: Psychic Traces and Somatic States

Let us go back to the process whereby a perception is inscribed. As we saw in chapter 5, a perception of the external world leaves a trace that we have called the sign of perception or the signifier. In a later process another perception can leave another trace and hence another sign of perception, that is, another signifier. Still later these two primary traces can become associated, leading to a new trace produced by the effect of this association. From this rearrangement a new signifier will result. To return to our diagram (see fig. 5.4 in chapter 5), a perception 1 leaves a trace 1 or signifier 1, perception 1 being the signified 1; a perception 2 (signified 2) leaves a trace 2 or signifier 2, and so forth. After this process an association may be established between trace 1 and trace 2, hence between two initial signifiers (signifiers 1 and 2), producing a new trace A, that is, a new signifier.

What we are suggesting is that this new trace resulting from the association of the two initial traces creates distance from the initial perception (of the signified) and that, by this process of transcription, the signifier newly constituted from the two initial signifiers is no longer in direct relation with the signified corresponding to external reality. This process would continue, from association to association, to form, for example, trace X or signifier X.

How can we link the somatic state to this new set of data (fig. 7.1)? If it is true that a given perception, leaving a given trace, can be associated with a given somatic state—let us say that perception 2, which leaves trace 2, is associated with a somatic state S—and that this trace is associated with others to constitute new traces leading to trace X, hence to a signifier that is no longer in any relation to the initial signified, we come to the conclusion that somatic state S, which was originally associated with trace 2, is now associated with newly constituted traces up to trace X, which becomes XS, this last constituting one of the elements in an unconscious fantasy scenario. We also infer from this that the somatic state marks traces that are unconscious from the outset (trace 3). With all these mechanisms taken together, the somatic state is thus carried as a marker all along the associative chain that leads to one of the elements of unconscious internal reality. The somatic state is therefore associated at the end of the chain with a trace that no longer has any direct relation to the trace coming from external reality.

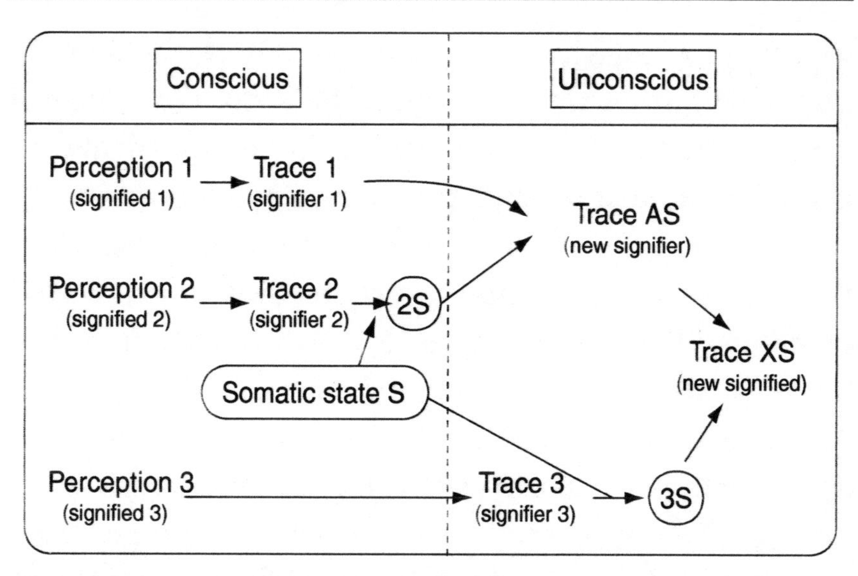

Figure 7.1 This diagram suggests that a given somatic state S may become associated with an initial conscious trace (trace 2). This somatic marking of the primary trace is conveyed all along the signifying chain and becomes associated with an unconscious trace (trace XS) that result from successive transcriptions and associations. We may imagine that the same phenomenon comes into play with traces that are unconscious from the outset (like trace 3 in our diagram, which thus becomes trace 3S).

Let us take a simple example to illustrate these associations between somatic state and perception. When an infant is under the pressure of thirst and hunger it experiences a state of marked somatic distress that can be calmed by the maternal breast (which Freud understood to be the primal experience of satisfaction). From the somatic point of view, the levels of glucose are low (hypoglycemia), for the reserves of energy have been used up. The infant is thirsty, which from a biological perspective corresponds to a hyperosmolarity of its plasma; that is, the concentration of salt in its blood is higher than in a physiological situation: the child is dehydrated. These biological variables, glycemia and osmolarity, are detected by the brain in the hypothalamus, where specialized neurons, sensitive to glycemia and osmolarity respectively, are activated (Koizumi 1996). As we have seen, at each moment the brain "reads" the state of the body.

Thus we are in a precise, objective somatic state characterized by plasmatic hypoglycemia and hyperosmolarity detected by the brain. This disturbance of homeostasis may correspond to the infant's state of distress as defined by Freud (1926). The infant experiences this tension as unpleasure, manifested, for example, by crying. But this crying is perceived by the child, who both produces it and hears it. What we are dealing with, ultimately, is a perception, a perception of external reality. We are in a prototypical situation of a somatic state (hypoglycemia, hyperosmolarity) associated with a perception of the external world (crying).

The other responds to this crying.[1] The mother approaches the infant and presents her breast, a source of glucose, other energy substrates, and liquid. This act, offered in the simultaneity of the insistence of unpleasure linked to a given somatic state, reestablishes rather

1. This other, called the *Nebenmensch* by Freud, is the external source of help offering the specific action of someone well informed, to paraphrase Freud (1895) in his remarks on the experience of satisfaction.

quickly the physiological values of glycemia and osmolarity. The concatenation produces a discharge of the tension connected with this somatic state of unpleasure (fig. 7.2). Unpleasure is followed by pleasure: this is the famous experience of satisfaction described by Freud in *Project for a Scientific Psychology* (1895). As a result of the specific action of the other (realized in simultaneity, *Gleichzeitigkeit*), the infant passes from a state of distress

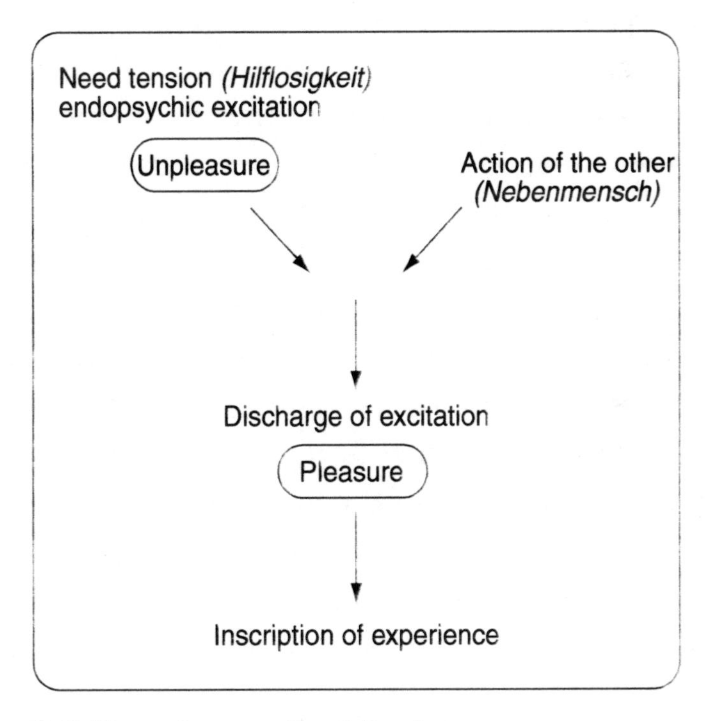

Figure 7.2 Experience of satisfaction.
The unpleasure produced by internal excitation linked, for example, to need tension plunges the infant into a state of distress. Only the other (*Nebenmensch*), through a specific action produced in simultaneity (*Gleichzeitigkeit*), makes possible the discharge of excitation and the experience of satisfaction, the result being the inscription of the experience.

of somatic origin to a state of well-being that Freud associates with the concept of pleasure. It is on this basis that experience is inscribed by means of the mechanisms of plasticity.

We might say that the child's somatic state generates a vital drive that ends in crying. This crying is itself a cry of the living being, at this stage without mental intentionality and hence without particular meaning. But it triggers the action of the other that brings soothing. In this way a first simultaneous association[2] arises among three perceptions: the perception of the dysregulated somatic state; the perception of the crying; and, finally, the perception of the soothing that comes from the warm, comforting liquid from the maternal breast.

The mechanisms of plasticity create a trace that becomes a mnemic trace, unconscious at this stage but definitely present. To return to the terms we used above, a new trace is constituted by the association of two primary signifiers, in our example the sign of the perception of the crying and the sign of the perception of the presentation of the breast. What is more, a given somatic state (S in fig. 7.1) is associated with the perception connected to the crying.

---

2. The specific action realized in virtually simultaneous fashion—in *Gleichzeitigkeit*—is, for Freud, the basis of all connections among neurons. Lacan, too, emphasizes the fundamental aspect of this synchrony: "Everything begins when several signifiers can present themselves to the subject at the same time, in a *Gleichzeitigkeit*" (1959–1960, 65). In this simultaneity we once again find the coincidence of synaptic trace, psychic trace, and signifier discussed in chapter 5.

If we look at this from the perspective of Freud's energy theory, we can say the somatic state triggers a vital drive pushing for the discharge of this unpleasant somatic state. The external object, mother's breast, makes this discharge possible. It is thus the action of the other that permits the discharge, leading to the satisfaction of the vital drive. A somatic state, hence perceptions coming from the internal world, is associated with events coming from the external world.

Here we have a simple module relating a somatic state, a perception, and a discharge of a drive linked to the somatic state. This discharge occurs through the other, specifically through an object, the breast. What is the status of this object in the chain of inscriptions? Through the response of the other the breast is associated with the crying. As a result, what is associated with the crying, for example the passage from the somatic state of distress to a state of calmness, is thus associated with the breast. What is the status of this breast? In the first stages of the establishment of this network of associations the breast is actually a concrete object made of flesh. The processes of transcription detach it from its initial state. A trace of experience is substituted for the experience proper: a mnemic trace. Thus we have, on the one hand, the object involved in the experience and, on the other hand, the trace of the experience, which involves the object but in the form of a representation, that is, in the literal sense, a presence of the object in its absence. In Freud's sense this would be a perception without an object, which he calls a hallucination of the object.

From then on the object can be only a placeholder for the first object that was associated with the inscription of the experience. The object has been split into this first object and the object that then comes into play through a representation. Our model illustrates the necessary distinction between the object that becomes a representation and the thing corresponding to the real object of the original experience.

In the first phase of the association between a somatic state and an object that allows satisfaction the object and the thing are identical, as in the pair *perception 4/trace 4*. Here, on the other hand, the object does not correspond to a representation; it is something real, the breast, that is *das Ding* according to Lacan in his reading of Freud (1959–1960, 101–114).

Once these initial associations are well established (hunger/thirst/crying/breast/discharge), we can imagine other events. For example, the child is in his room, hungry and thirsty. He cries, and the mother, hearing the cry, opens the door. In opening or closing, the door makes a particular sound. Little by little the infant creates a second association between the sound of the door and the maternal breast that will appear several seconds later. There is thus a new association being created between the crying and the sound of the door opening or closing to bring discharge very soon thereafter. Here object and thing are dissociated from the outset, that is, the object bringing satisfaction is no longer the thing, the original breast. The object here is already at a distance. The sound of the door has become the representative of what will make satisfaction possible. The thing remains the breast, but in the child's psychic construct an initial dissociation has been made between the object and the thing. The object associated with the experience of satisfaction is materially distant from the thing; the sound of the door, after all, is quite different from the mother's breast. But the two perceptions are bound by the mechanisms of synaptic plasticity into a new signifier that can be used for the experience of satisfaction.

Let us go a little further. The child grows older and no longer needs the maternal breast; he can feed himself more or less alone. Yet the mother's presence remains a pleasant experience in the face of potentially unpleasant somatic states, as when he cries, feels lonely, or has hurt himself. The mother's presence is a source of soothing, that is, it makes the unpleasure stop. Unfortunately, this also implies the reverse: the association can be so strong that the absence of the mother becomes a source of unpleasure and anxiety probably associated with a somatic state of distress. Left alone, the child will search for soothing, possibly using his own body, for example by sucking his thumb compulsively. In themselves these actions are associated with somatic states perceived as pleasant. Discharge is sometimes even sought through emerging sexuality, especially infantile masturbation, which ends in extremely strong perceptions that are undoubtedly associated with very marked neuroendocrine changes.

This serial effect can obviously become infinitely complex. If the mother, for example, often wears a pink shirt when she comes to reassure the child, this pink shirt gradually becomes associated with an experience of pleasure, because unpleasure ceases, and a new association will be created, this time between pink shirt and experience of satisfaction. When he is a little older, the child will discover sexuality with a nanny who also often wears a pink shirt. This time the experience of satisfaction will clearly be of a sexual nature; it will find its realization in the ancillary sexual experience in which pink shirt and experience of satisfaction are associated. Thus, in the course of time, a complex series of representations is formed in association with a given somatic state, but this series can be followed back to the initial module of the primary experience of satisfaction in which the first link was forged between a representation and this specific somatic state.

8

Man and Wolf: Fantasy, Object, and Action

Let us imagine a given somatic state that, in the past, was associated with a rather unpleasant situation. The activation of this somatic state induces an unpleasant internal perception that makes the person try to produce strategies to get free of it, or, to remain with our original module, to reestablish a homeostasis. The unpleasure induced by the somatic state leads to a discharge.

In the case of the infant, this discharge comes about through the specific action of the other. In the case of the adult, it can come from the person himself. Still remaining with the logic of the original module, we see that the action discharging the state of distress involves the outer world; the discharge does not occur in a vacuum but requires an object from external reality, the object being considered in the broadest sense of the term. The goal that consists in producing a discharge is predominant. In contrast, the nature of the object is more or less indifferent as long as it can function to permit the discharge of excitation. Discussion of this point calls for reference to Freud's drive theory, and this will be the focus of the following chapter. For the moment let us be content to recall that the object is what is most variable in the drive[1]: it may be any object whatsoever on condition that it can serve to bring about discharge and produce satisfaction. Fantasy is precisely a setup that binds the person to the misrecognized object of the drive[2] and can place any other object in series with this enigmatic object through a specific unconscious scenario.

Seen in this light, the somatic state then triggers a drive that must find an object in order to be discharged. This object may be very far from the thing (*das Ding*), the first object involved in satisfaction. It is only a placeholder for an original object that has been lost but remains in-

1. "The object . . . of an instinct is the thing in regard to which the instinct is able to achieve its aim. It is what is most variable about an instinct . . . . It may be changed any number of times in the course of the vicissitudes which the instinct undergoes during its existence . . . ." (Freud 1915a, 122–123). For further discussion of Freud's drive theory in connection with progress in present-day neuroscience, see chapter 9. (Translator's note: For the purposes of this book, the term *instinct* used in the standard English version of Freud's works may be considered equivalent to what is here referred to as *drive.*)

2. This is illustrated by Lacan's (1955–1956) matheme of fantasy, in which the barred subject (representing the division of the subject) is bound to the misrecognized object of the drive. On the relation among the object of the drive, the object of fantasy, and the object of desire see Lacan 1964.

scribed in the form of absence and unconsciously orients the action because it is already at work in fantasy life. The object of the fantasy is peculiar to the person's history. It functions for one person and not necessarily for another, linked as it is to a given individual through his fantasy scenario. In this way somatic state, drive, object, and fantasy are linked in a cycle that can be activated from without but also from within if the fantasy is indirectly placed in tension through objects or situations that are able to evoke it. The activation of the fantasy then generates an unpleasant somatic state that, via the drive, will be discharged in an action involving an object.

We thus find ourselves in a restrictive loop that brings into play body, drive, object, and fantasy in a cycle that can be repeatedly, and sometimes unexpectedly, activated. We can see the extent to which the object cannot be taken in the literal sense of the word. It must be conceived as participating in a situation, behavior, or symptom organized by the fantasy peculiar to the individual (fig. 8.1).

Let us imagine the fictional case of a couple who have just formed a relationship. Each of the two is still in contact with his or her previous lover. Despite the passion they feel, there is not yet room for their new history together. They are tormented, torn by the irruption of what took place before they met. Yesterday, as it happens, the man got a letter from his former girlfriend, which led to a violent quarrel: she reproaches him; he defends himself, swearing that it's all over, that she is the one he loves. She pushes matters very far, won't listen to him,

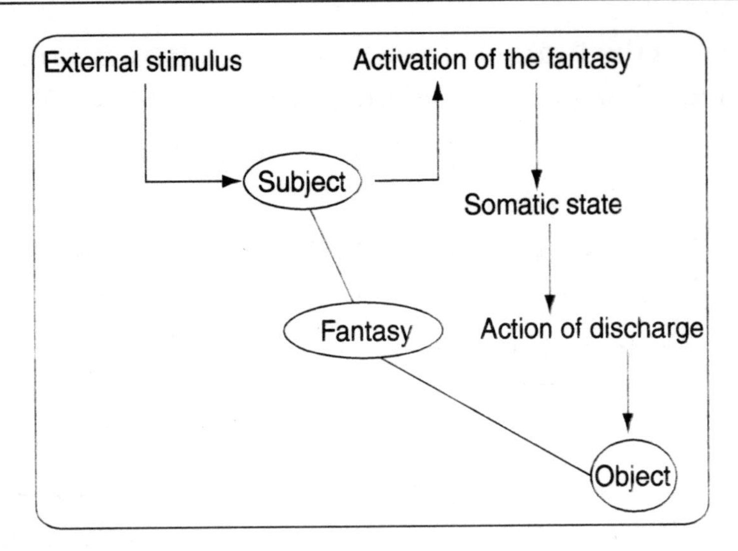

Figure 8.1 A stimulus from external reality activates the fantasy, which, being associated with a somatic state of distress, can be resolved by an action that introduces an object. The object is bound to the subject by the mediation of the fantasy. In this diagram we see how the subject and the object are linked by the fantasy.

rejects him. The next day she is the one who gets a letter from the man she used to be with, which provokes the indignation of her new lover. How could she have gone on the attack like that when she clearly has the same ambiguous relationship with the man she left? The quarrel picks up where it had left off and becomes twice as intense. He feels betrayed. He's in despair. The anger escalates to the point where it blinds him. He no longer knows what he's doing. He no longer knows who he is. He is overcome by immemorial distress like the kind he so often experienced as a child after he was abandoned and placed in an institution that rejected him and left him in a state of extreme loneliness. He is no longer in the present situation but has once again found what has tormented him ever since childhood. He no longer knows who he is: he is again seized by this utter distress he feels each time he faces the risk of aloneness.

For him, if she is in contact with someone else, he himself doesn't exist anymore. This is his fantasy: if he is abandoned, he no longer exists. This conviction overwhelms him. Absence torments him, persecutes him. Absent from himself, he has to get a grip on himself, save himself. And so he hits her, once, twice. Loving passion has turned into murderous passion, and this is how passion can kill without intending to. He hits her without knowing what he's doing, as if to save himself and stop being in this distress he has fallen back into. He hears her cries, but they seem unreal. She collapses, falls at his side. He loves her. He wants to be with her. But she does not move. Because he could not be separated he wound up destroying her to save himself. He lost her in order to find her again, to find himself again. He is present now. He sees the effects of his acts. This violent discharge has saved him, but she is the one who isn't there anymore.

Let us review what happened for this man who discovers the presence of another man in his girlfriend's life. This discovery, however painful, could be dealt with rationally, but in the present case it directly activates a fantasy scenario constructed around jealousy, or, more precisely, around the idea that, if she is in contact with someone else, he himself does not exist. He experiences a violent sense of devaluation, indeed of negation of his own individuality. The activation of this fantasy is associated with an unbearable somatic state connected to the distress he felt at an early age when he was placed in an institution. This unbearable somatic state leads to an action determined by that particular experience instead of by the present situation. In our fictional case the recollection of childhood distress ends in a discharge of the tension linked to the somatic state in the form of a violent act that ultimately suppresses the other who had taken the place of the absent, lost object. This discharge could also have gone in the direction of self-destructive violence to the point of suicide.

Let us leave this couple with their destructive passion and return to the question of the discharge of the drive linked to the activation of a somatic state. Luckily there are other, less dramatic, modes of discharge. Whatever the form taken for the discharge of excitation, we must understand that the activation of the somatic state is triggered by the activation of a fantasy scenario inscribed in unconscious internal reality. The theory of somatic markers (Damasio 1994) provides a biological substrate for the puzzling question of the somatic anchoring of the drive.

Although the fantasy scenario is peculiar to each person, it can be aligned with a limited range of prototypical fantasies (primal scene, seduction, castration, etc.) that can never be resolved by the small child confronted by them in the enigmatic reality of his body and its surroundings. From this point of view there is no need for a traumatic event in reality (abandonment, violence, etc.). A child may be marked by life itself, whatever it may be: by his body, whose integrity he is unsure of; by his parents, who are connected by something he does not know; by his place in the desire of the other, who may also constitute a destabilizing enigma for him. By way of analogy we might say that fantasy scenarios are like literary genres: though they are limited in number, their contents are different and unique in each case.

Now once a fantasy is activated, it connects up with a specific somatic state, demanding a discharge that short-circuits all reason. This is how the phenomenon of violence is to be understood. Yet we must distinguish between destructive violence and the kind that saves. Thus, paradoxically, violence can come into play to save us from states of distress. The Greeks spoke of two complementary powers: *Eris,* which is quarrel, discord within what is united, and *Eros,* the union of what is dissimilar (Vernant 1999). Eris and Eros were inseparable for them. We find the pair included in the idea of violence, which contains these two contradictory forces. For on all levels violence entails opposite tendencies: a violence of life, which is form-giving, constitutive, and saving, and a violence of death, which is destructive and leads to suicide, murder, abuse, racism, and genocide. When a person produces a violent, destructive discharge, he does so in order to save himself from a destructive state of distress.

The word *violence* is in itself contradictory. On the one hand it is related to the word *violation* in the sense of rape, to the idea of breaking and entering, domination, negation of otherness; on the other hand it is also related to the idea of vigor, potency, vital force (Héritier 1996). In addition, we must distinguish between violence undergone and violence exercised, and we must remember that violence can save, that it is twofold: it can be loving and legitimate, and it can be impetuous and tyrannical. These are the terms Pascal used when he wrote that a child snatched by his mother from the arms of kidnappers must, in his distress, cherish the loving, legitimate violence by which she procures his freedom and must hate only the impetuous and tyrannical violence of the men who are holding him unjustly (see Nancy 2003). Using the word *violence* in describing a situation does not distinguish between these different dimensions reflected in the ambiguity of a word that, in itself, does not tell us on which side we stand.

This contradiction inherent in violence is the sign of a possible connection between the forces of life and the forces of death. There may even be, paradoxically, an aspect of life in the destructive violence that is sometimes mobilized to save an endangered identity. A person may become violent in order to save himself, as he may do in order to save someone else. Thus there is not only a contrast between life violence and death violence but also a division within all violence between a life drive and a destructive drive, as Freud (1920) put it in his theory of drive dualism. If there is an interweaving of these drives, we are on the side of life, even if it is in its aggressive valences. If these valences are disentangled, on the other hand, we are on the side of destruction, under the absolute reign of Thanatos, as in our fictional case.

The duality between life and destruction thus seems intrinsic to the process of violence, independently of its aim. Even if we can distinguish between a violence of life and a violence of death with regard to the intended aim—we cannot equate the violence of rape and the violence of desire, the violence of destruction and the legitimate or heroic rage of the fighter—the violence that destroys is not necessarily different in its mechanics from the kind that aims at life. As soon as violence comes into play, one and the same process has been begun, a process involving a drive that is discharged in an action making use of an object.

All violence is a breaking and entering. It denatures what it aggresses against and what justifies it. Violence remains above and beyond the reasons invoked to justify it. It somehow traps us in a fantasy that is outside external reality.

It might be argued that violence is inscribed in the animal nature of man, as in the saying that man is a wolf to man (Freud 1929). In any extreme situation in which violence erupts, it is the animal part of man that will be expressed, it is said, for violence is a remnant of evolution that seizes the person beyond his humanity.

This claim can be refuted. Ethologists themselves have shown that wolves do not treat each other as destructively as men do. Without going further into the discussion, let us note the philosophical tale told by

Baltasar Gracián (1651–1657), who poses a serious challenge to the adage *homo homini lupus* (man is a wolf to man).

Gracián tells the story of Critilo's reunion with his son, Adrenio, whom he had thought lost in a shipwreck. The boy was found on the shores of an island where he had grown up all alone among animals. Adrenio is a wild child who cannot speak, and Critilo teaches him the use of language, which he could not have gotten from the beasts. Though Adrenio is glad to encounter human society, Critilo warns him against men, presenting them as more dangerous and more destructive than all the animals taken together. Each man is a wolf to his fellow men, he says, and men have even taught tigers to be more cruel than they were by nature. According to Lacan (1948), Gracián teaches us that man's ferocity is greater by far than that of the animals, even the carnivores.

But if it is true that no animal surpasses man in cruelty, is it really man's animality that is revealed in violence? Isn't acting violent a distinctively human trait, as Freud observes in *Civilization and Its Discontents*? People's neighbor, he says, "is someone who tempts them to satisfy their aggressiveness on him, to exploit his capacity for work without compensation, to use him sexually without his consent, to seize his possessions, to humiliate him, to cause him pain, to torture and to kill him" (1929, 129).

If it is in man's nature to behave this way, what do we understand by "human nature"? Is it an innate, biologically inscribed dimension that would appear on its own, ultimately without the person's knowledge? Isn't it better to favor a purely subjective view of the phenomenon? To be sure, the mechanics of violence involve the body, but what sets it going really does seem to be connected to the person's mental life, to what has determined him in his history and orients his action. Such a concept of violence entails a conflictuality that is peculiar to the individual and mobilizes his internal drivenness.

Yet seeing violence as having to do with the drives instead of as a hypothetical instinctual animality does not mean that we can do without the dimension of the body. Violence as a drive phenomenon involves both the body and the subject, the tension between a fantasy scenario and an associated somatic state calling for discharge.[3] Violence cannot be conceptualized without the body, but this is quite different from imagining it as an animal, biological phenomenon apart from the psychic.

Here, then, is where we find the relation between the secondary traces constituting the fantasy scenario and the associated somatic states. Thus we can say that the drive, the product of the association of a fantasy with a somatic state, involves a discharge, for example the irrepressible triggering of the violent act in our example. It would not be amiss here to connect this concatenation to the decision-making processes and their somatic anchoring as elegantly formulated by Antonio Damasio (1994), who defines decisional processes as coming primarily from the conscious world, either directly through

3. Let us recall that Freud defines the drive "as a concept on the frontier between the mental and the somatic" (1915a, 121–122).

perception or through representations accessible to consciousness. In his view, making a decision and acting on it are determined by the anticipation of a somatic state intended by the action.

This approach is above all concerned with the conscious, cognitive level, but we believe it might be possible to imagine the same type of process for the fantasy scenario inscribed in unconscious internal reality and the drive dictated by the somatic state associated with this fantasy. Considering the brain as an organ among others, one that is able to "read" the somatic state and represent it to itself, along with directing action, we end up with a logic of action as determined by the drive discharge produced in interface with the fantasy scenario and a somatic state.

The central point recurring at each of the turns of the spiral of this book is that behavior is determined by the perception of external reality and by an unconscious internal reality that interferes with this perception. The two perceptions—of external and internal reality—are associated with particular somatic states. Perceptions coming from unconscious internal reality, that is, those connected with the activation of fantasy, are associated with somatic states perceived very strongly by the individual. It is in this way that fantasy intrusively interferes with the perception of external reality and determines the act that is produced, which may be very far from what might have been the motor response in direct connection with the external stimulus.

9

An Unexpected Phone Call: How Drives Originate and What Becomes of Them

To simplify, we can say that if unconscious internal reality and its fantasy scenarios did not exist, a stimulus coming from the external world would lead to an action in direct relation to it (Action 1, fig. 9.1). Let us take an example drawn from the arsenal of neuropsychological testing. The instruction is to recognize in a series of images those that represent animals. Intermixed with the series of animal images are images having nothing to do with the animal world. The subject has to press a button when an animal is present. With a normal subject there will be conformity between the instruction and the response; that is, when an animal is presented the subject will press the button.

This test actually explores a whole range of levels in the treatment of the external stimulus, from visual recognition to the executive functions resulting in action, passing through the activation of associative regions and

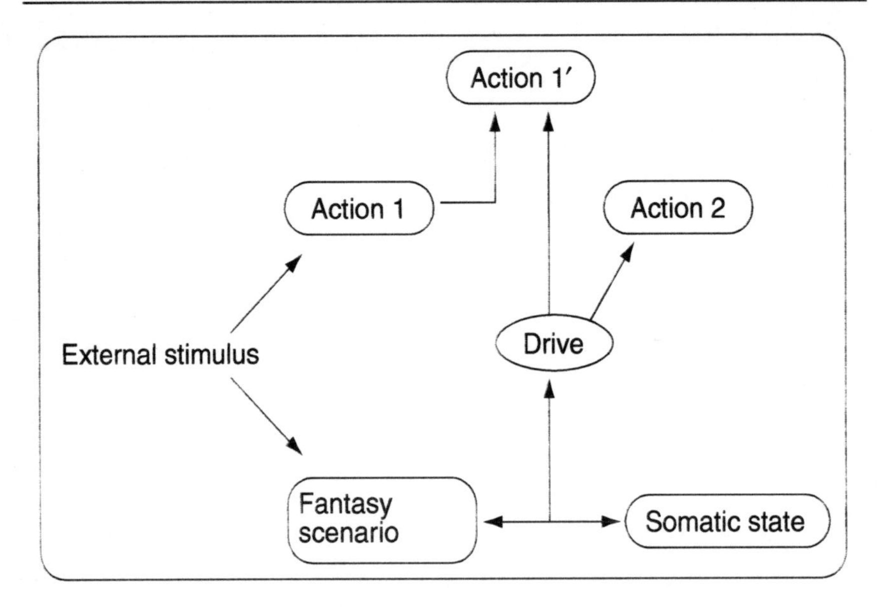

Figure 9.1 This diagram emphasizes the fact that the drive is at the interface between the psychic (unconscious fantasy scenario) and the somatic (somatic state) as well as the fact that the drive discharge can interfere with action.

the correlation of the perception with the images stored in the memory systems. Thus there is conformity between perception and response in this specific context.

In our example the mobilization of unconscious internal reality introduces an incompatibility between external stimulus and action. We may imagine that, in the wake of a process mobilizing unconscious internal reality, the subject will occasionally fail to recognize certain animals presented. This absence of recognition on the part of a normal subject could be ascribed to the margin of error acceptable in this kind of test, and this is done routinely. Nevertheless, at least as a matter of speculation for the moment, we might see here the effect of an intrusion of unconscious internal reality on the model of what is called in psychoanalytic terms the formations of the unconscious (parapraxis [bungled act], slip of the tongue, lapse of memory). What is identified as mere error on the conscious level thus takes on meaning in an unconscious theater.

Let us return to this unconscious reality. We have posited that it is constituted by the mechanisms of plasticity described above. What characterizes this unconscious internal reality is that it can be activated by stimuli coming from external reality and that it is associated with a particular somatic state. The result of the association between the fantasy scenario and the somatic state is a tension corresponding to the drive as defined by psychoanalysis.

According to Freud (1915a), the drive is located at the interface of the psychic and the somatic; it represents on the psychic level the excitations from within the body. The tension linked to the somatic state is perceived as unpleasant, for it disturbs the homeostasis of physiological states. This tension should be able to be discharged, either by an action in direct relation to the disturbed state or by an action determined by what is inscribed in the fantasy scenario. Thus, to discharge the tension linked to this somatic state, the person can mobilize an action with no direct relation to the initial stimulus (Action 2, fig. 9.1). The drive may also interfere with an action in direct relation with the external stimulus (Action 1) and make it veer off toward an unforeseen action (Action 1′) that may surprise the subject, as in the psychoanalytic model of the parapraxis.

Let us take the example of a man who wants to get in touch with his wife quickly by telephone. He mechanically dials the number he knows so well. To his great surprise, another woman answers. Not just any woman, but the one with whom he once had a relationship he thought was over. He recognizes her after a brief pause; one always remembers clearly how voices sound on the phone. He can't believe his ears: what happened to him? Everything had been dealt with. But hadn't he had a daydream of seeing her again soon? And now here he is, doing something in spite of himself: Action 1 has become Action 1′. And Action 1′ is the one that corresponds to a wish he had relegated to the past. It isn't so easy to escape the short circuits of the unconscious that, in this case, are traveling over the telephone wires.

Thus action is not solely the result of the influence of external reality. It may also be generated from unconscious internal reality, which comprises a fantasy construct activated unbeknownst to the person and determines an unconscious wish that exerts pressure. What seems to be an inappropriate action may turn out to be an appropriate one with regard to the unconscious wish. Thus an action may be directly triggered by unconscious internal reality without the intervention of an external stimulus. It may be produced from an unconscious fantasy scenario in resonance with a somatic state. The drive emerges from this interface. Likewise, the action may be generated from the activation of a somatic state that evokes a fantasy scenario, and this, too, leads to drive discharge (fig. 9.2).

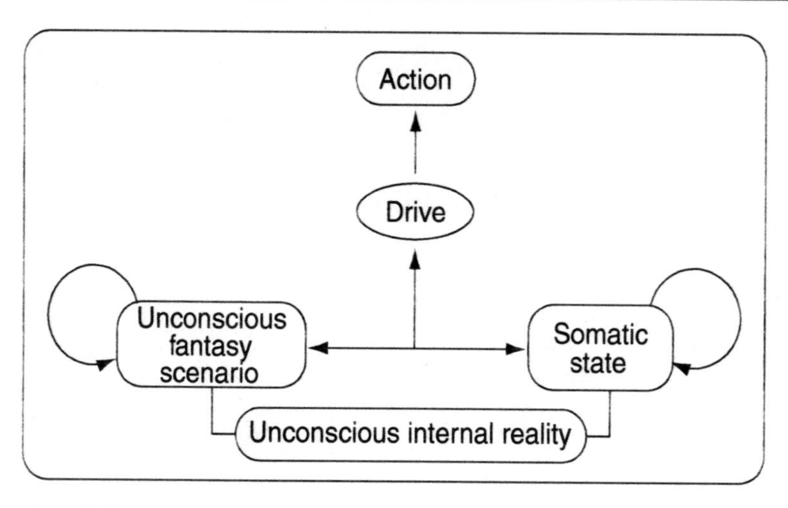

Figure 9.2 Origin of the drive from unconscious internal reality, whether through the activation of the fantasy scenario or through the activation of the somatic state.

Finally, there is a third diagram indicating interference with an action corresponding to an external stimulus (Action 1 in fig. 9.3), from a solely endopsychic excitation of the fantasy scenario, without it being necessary for this scenario to be directly activated by the external stimulus.

If the drive coming from unconscious internal reality imposes its law on the action, what we have to do here is take a closer look at the features of this other reality. Unconscious internal reality has no dimensions. Temporal reference points of past and present overlap. Places become confused with each other. A thing and its opposite can coexist without contradiction. Elements are associated without regard to either contradiction or negation (to repeat: fantasy is fundamentally without dimensions). The person can be himself and someone

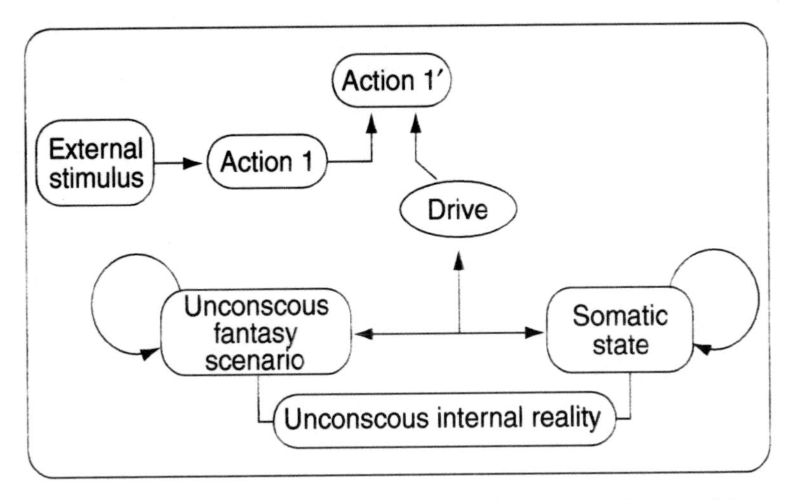

Figure 9.3 Interference with action by an endopsychic activation of unconscious internal reality

else at the same time: male and female, here and elsewhere, in the past and in the future. The signifiers of fantasy do not correspond to the signifiers of the external world and concrete reality.

Thus it would come as no surprise if there were no compatibility between stimulus and action to the extent that the stimulus comes from unconscious internal reality; the laws of the unconscious have come into play, those of the primary process defined by Freud.[1] The fantasy and the associated somatic state also generate stimuli—the drive—calling for an action to resolve the tension

1. "To sum up: *exemption from mutual contradiction, primary process* (mobility of cathexes), *timelessness,* and *replacement of external by psychical reality*—these are the characteristics which we may expect to find in processes belonging to the system Ucs [Unconscious]" (1915b, 187; emphasis in original).

they entail (figs. 9.2 and 9.3). Thus the drive is held to be the result of the association of the fantasy with the somatic state. Everything tends toward a discharge in order to reestablish homeostasis.

Let us go right back to the way Freud defines the drive:

> If now we apply ourselves to considering mental life from a *biological* point of view, an 'instinct' appears to us as a concept on the frontier between the mental and the somatic, as the psychical representative of the stimuli originating from within the organism and reaching the mind, as a measure of the demand made upon the mind for work in consequence of its connection to the body (1915a, 122–123; emphasis in the original).

Freud defines the drive in terms of its source, its pressure, its aim, and its object. The source is "the somatic process which occurs in an organ or part of the body and whose stimulus is represented in mental life by an instinct [i.e., a drive]" (1915a, 123). The source of the drive is thus the Freudian version of what we have called the somatic state. Freud already acknowledged that "[T]he study of the sources of instinct lies outside the scope of psychology" (123). Though the fact of its coming from the body is, for Freud, the absolutely determinative element for the drive, all we know of the drive in mental life is its aims. The aim of a drive is always its satisfaction, which can be obtained only by suppressing the state of excitation at the source of the drive. The pressure of the drive, Freud tells us, is its motor factor. As for the object, it is simply "the thing in regard to which or through which the instinct is able to achieve its aim" (122).

The object is thus the most variable aspect of the drive. It is linked to the person in fantasy without being originally linked to the drive (see fig. 8.1 in chapter 8). It is merely what makes satisfaction possible and may be "changed any number of times in the course of the vicissitudes which the [drive] undergoes during its existence" (1915a, 122–123).

The drive has its origin in the association between a fantasy scenario and a somatic state; it requires for its discharge an action and an object that are not necessarily in relation with external reality. This raises the question of the object of the wish, which can be identified from the object of the drive, hence from a demand coming from unconscious internal reality. Yet there happens to be a confusion between the object of the wish coming from the drive and the object of the wish identified by the person on the basis of his conscious cognitive life. We see this, for example, in certain situations of social success in which an object—an objective—is identified as the ultimate object of desire.* Yet, paradoxically, obtaining this object may reveal the unattainability of the

*Translator's note: French psychoanalysts use the term *desire* (*désir*) where English translations of Freud's term *Wunsch* use *wish.* I am using *desire* where the context is Lacanian in orientation, but, as in the case of *instinct* and *drive,* the terms *wish* and *desire* should be considered equivalent for the authors' purposes.

object of desire. This realization can lead to despair, sometimes even to suicide following the achievement of a goal the person had set for himself. On the conscious, cognitive level the person thinks he has identified an object of desire. He organizes his life, his career, so as to attain it. He attains it but finds himself plunged into an urgent sense of dissatisfaction. He has somehow missed out on the object of his desire, which in fact came from his fantasy.

For fantasy is what supports desire (Lacan 1962–1963). Hence the object of the fantasy, or, more precisely, the misrecognized object of the drive, turns out to function as the object that causes desire, not the object to be obtained as the goal of desire.[2] Thus awareness of this discrepancy can unexpectedly plunge a person into despair. Present-day consumer society constantly bombards us with preconditioned objects of desire, one and the same for everybody. A person's energy and skills are invested in obtaining gadget objects that leave him unsatisfied because they have nothing to do with the object causing his desire. There is an irreducible gap between the object causing unconscious desire and the objects imposed by society. The object of desire is not the object of alleged need imposed by the market. It rises instead from the distance between need and the demand that always aims at something beyond the object.

2. The object cause of desire (Lacan's *object a*), that is, the conditionality of desire, is not the same thing as the object at which the desire is aimed, its intentionality. Jacques-Alain Miller's discussion on June 2, 2004 of this aspect of Lacan's seminar on anxiety (1962–1963) is as yet unpublished in English.

Drive energy does not always find release. It can turn back against the person himself, stimulating unconscious internal reality and heightening unpleasure, sometimes in a total inhibition of action. This is what we find in neurotic phenomena, where it is well known that the person is not always operating for his own good. He may even produce actions that, once they are perceived, become in turn a stimulus reactivating the fantasy in a self-maintained vicious circle. The drive movement originating internally can be experienced as coming from the outside (fig. 9.4). The person is being acted on by himself without being aware of it. We could call this a transduction, a transportation, a transference in the true sense

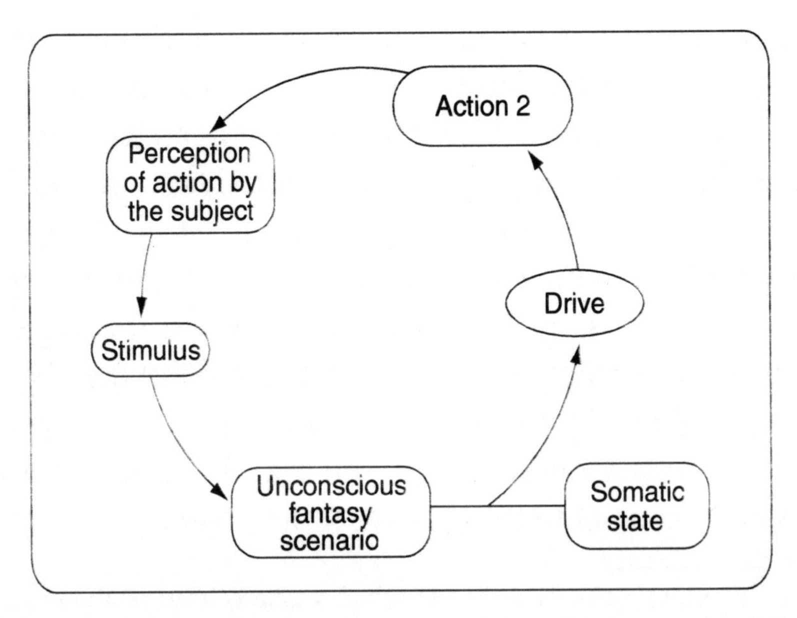

Figure 9.4 Action 2, produced by drive discharge, itself becomes a stimulus experienced as external, activating the fantasy scenario, consolidating it, lending it increasing reality, until it appears like an external reality.

of the term,[3] of the drive (internal) into a stimulus (external). In this circular modality we see that fantasy can go together with the somatic state as the origin of the drive just as well as it can become one of its destinies. Once the fantasy scenario is reactivated in this way, the somatic state is reawakened, producing a need for drive discharge that leads to a limited and limiting action that maintains the vicious circle. This is a classic mechanism of neurotic concatenation.[4]

Psychoanalytic work consists precisely in connecting the fantasy scenario with a congruent somatic state that may be associated with it. When these two elements are linked, the interfering action that results from their separation disappears. In the ideal case the patient returns to a situation in which the external stimulus triggers an action appropriate to the nature and intensity of that stimulus. He is once again on the path of consistent action, action no longer interfered with by fantasy. After a psychoanalysis the fantasy continues to exist, of course, but, to the extent that the patient has been able to experience its type of connection to a somatic state, he can

3. Use of the term *transference* in this context is not unwarranted. Lacan himself spoke of transference as "the enactment of the reality of the unconscious" (1964, 146).

4. All this is considered to take place beyond the pleasure principle, maintaining a *jouissance* of the fantasy that keeps the person under the pressure of an unpleasure he seeks, despite himself, in a repetition compulsion that is masochistic in aim. See Freud 1920 in this connection. Translator's note: on the concept of *jouissance* see chapter 15 below, p. 236.

rely on the fantasy, making use of it to find the path of an action consistent with reality. He is no longer caught by surprise, to his cost, by the irruption of an action determined by fantasy. Having been able to tackle the fantasy in an analysis, he can orient himself in a new way and, at last, find a relation to reality.

10

Incest and the Refrigerator: Pleasure and Unpleasure

We have seen how the disturbance of the physiological state is a fundamental aspect in the production of the drive. From the physiological perspective, it is legitimate to suppose that the break in homeostasis is perceived as unpleasant. And indeed, this rupture represents a potential danger to the integrity of the organism and must therefore be signaled in such a way as to mobilize mechanisms for reestablishing homeostasis. Hence it is reasonable to posit that the signal of the rupture of homeostasis is perceived as unpleasant rather than pleasant.

An analogy could be made with the systems of pain perception, which are activated by irruptions into somatic integrity (for example, a wound) or by pathological processes active at a precise place in the body. No one will challenge the statement that pain is an unpleasant sensation but one that has a protective function. The same would be true of the perception of the rupture of homeostasis, experienced as unpleasant, that is, felt as a state of unpleasure. It is interesting to note that the nerve fibers transmitting information on the somatic state, the interoceptive pathways mentioned in chapter 6, have the same electrophysiological properties (speed and mode of conduction of the nervous signal) as the fibers conducting pain, fibers of the type Aδ and C (Craig 2002).

Let us begin looking at the question of how homeostasis is reestablished. Physiology describes a whole series of negative feedbacks, that is, self-regulatory loops in which a physiological variable is maintained by the operation of regulatory systems like the endocrine system. Thus, after a large meal, the increase in the concentration of plasma glucose detected by, among others, the beta cells in the pancreas, leads to the release of insulin by these cells. This promotes the storage of glucose in the liver, which reestablishes glycemia. Insulin is therefore released into the plasma to reestablish homeostatic levels of blood glucose (Drews 1996).

According to psychoanalytic theory, a state of mental tension is associated with an unpleasure from which the person must be freed. As we have seen, he must free himself from the excitation produced by the activation of a somatic state associated with the unconscious representations constituting the fantasy. This association gives rise to the drive, whose aim is to be discharged so as to bring about the cessation of the state of unpleasure. Thus the drive is discharged in an action that will tend to reestablish homeostasis.

We can see how physiological reasoning and psychoanalytic reasoning, which *a priori* stem from incommensurate domains, come into direct connection in regard to homeostasis and unpleasure. The principle of homeostasis joins the pleasure principle beyond any analogy. Freud's comment, already cited several times, that the drive is located on the frontier between the somatic and the mental is much more than a mere hypothesis. It is a point of intersection between psychoanalysis and physiology. The two join without losing their distinctive natures.

On the basis of this logical concatenation, therefore, we can say that the disturbance of homeostasis causes a state of unpleasure that is resolved by the discharge of the drive. This discharge, which leads to the neutralization of the state of unpleasure, can in fact be seen as a mechanism leading to pleasure. Thus the Freudian pleasure principle, which aims at the reduction of a state of unpleasure, is primarily a principle of non-unpleasure. It is in the association of mental traces (constituting the elements of the fantasy scenario) with somatic states that the drive emerges and these two incommensurables—mental life and somatic life—are joined.

We also find the connection of the two incommensurables in the correspondence between the concept of homeostasis and what might be described as the non-unpleasure principle. As we shall see, two aspects of homeostasis, its maintenance and the action of discharge to reestablish it, correspond to the Freudian notions of the constancy principle and the inertia principle. As

Freud noted as early as "Project for a Scientific Psychology" (1895), unpleasure arises from the increase of endopsychic tension. Pleasure, he says, is associated with its discharge. The pleasure principle is in the service of a law of inertia that calls for the discharge of excitation. In physiological terms, we could say that the principle of non-unpleasure is in the service of a law of the maintenance of homeostasis that calls for the reestablishment of physiological variables. External excitation can be resolved by motor discharge, internal excitation by drive discharge. In both cases the aim is the restoration of homeostasis.

The drive originates in an association between a fantasy scenario and a somatic state. The somatic aspect is quantitative[1] to the extent that it can be reduced to biological values. The unconscious representations, inscriptions, and successive transcriptions of experience are qualitative in nature. The drive thus associates energic phenomena that are quantitative in nature and linked to the somatic state with qualities determined by the representations inscribed by the mechanisms of plasticity. As Freud said early on, "In this manner the quantitative processes . . . would reach consciousness, once more as qualities" (1895, 312).

---

1. "[W]e have certain knowledge of a trend in psychical life towards *avoiding unpleasure*. . . . *[U]npleasure* would have to be regarded as coinciding with a raising of the level of [quantity] or an increasing quantitative pressure . . . . Pleasure would be the sensation of discharge" (1895, 312; emphasis in original).

Freud thus described a mechanism through which quantity becomes quality. Through its connection with representations, internal somatic excitation constitutes one of the possible stimuli for the psychic apparatus. This brings us back to the question of the influencing event, or, in physiological terms, the question of the stimulus. We can see how the stimulus cannot be solely external. As Freud (1938) says, the body itself can replace the external world as the origin of the stimulus. Thus, as Lacan observes, Freud's "Project" can be seen as "the theory of a neuronic apparatus in relation to which the organism remains exterior, just as much as the outside world" (1959–1960, 47).

There are, accordingly, two kinds of experience that could come to be inscribed in the neural network (fig. 10.1): those coming from outside (perceptions received through the mediation of the sensory organs), and those

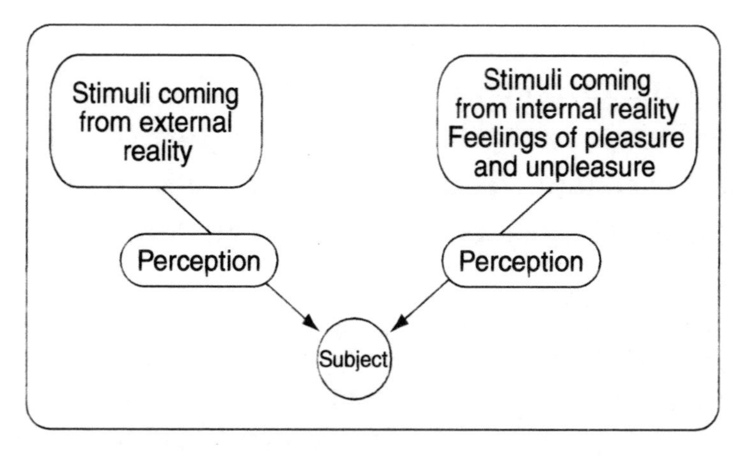

Figure 10.1 Stimuli may come from outside or inside. In this case, the body itself replaces external reality (Freud 1938).

coming from inside the body, detected by the interoceptive pathways (feelings of pleasure and unpleasure associated with the degree of internal energic tension, hence with a particular neurovegetative or neuroendocrine state).

Let us return to the physiological principle of maintenance of homeostasis, which Freud describes in terms of two principles, inertia and constancy. The inertia principle would come close to the physiological concept of reestablishment of homeostasis after homeostasis has been disturbed. As for the constancy principle, it seems to correspond to the foundational principle of physiology, namely, the maintenance of the homeostasis of the internal milieu (Bernard 1865). Here, too, we can see how Freud's reasoning converges with that of physiology. In Draft K of 1896, for example, Freud observes that the law of constancy is one of the "most fundamental attributes of the psychical mechanism" (1887–1902, 146). To sum up Freud's point of view: the inertia principle regulates the pleasure principle by maintaining as low as possible the quantity of excitation present in the psychic apparatus. This apparatus, however, is governed by another principle, that of constancy, that is, of the maintenance of a minimal level of excitation. Thus excitation must be reduced or, more precisely, maintained at a constant level.[2]

2. "The facts which have caused us to believe in the dominance of the pleasure principle in mental life also find expression in the hypothesis that the mental apparatus endeavours to keep the quantity of excitation present in it as low as possible or at least to keep it constant" (Freud 1920, 9).

The mental apparatus can thus be seen as an operator of homeostasis in which the somatic and mental aspects are joined. This statement, which may at first seem a bit general, is strongly confirmed by recent findings regarding the central role of the brain in the maintenance of somatic homeostasis. Here we must mention the role of various regions of the brain, especially the hypothalamus, the brain stem, and certain frontal regions (see chapter 13) in neuroendocrine feedback loops like that of the hypothalamus-hypophysis-adrenal axis, central in the mechanisms of stress and regulation of glucocorticoid levels, or the feedback mechanisms of satiety and the regulation of energy metabolism, to cite only a few examples (Jungermann and Barth 1996). Continuing with this line of reasoning, one could say that the drive leading to discharge, in principle behavioral in nature, is in fact an element in the feedback loop that helps maintain homeostasis in the organism.

The pleasure principle, then, is first of all an inertia principle that achieves a sort of automatic functioning (Lacan 1959–1960). In physiological terms, its process is similar to a negative feedback mechanism in a self-regulated system aimed at maintaining variables within physiological limits. It is striking to see how, according to Freud, the inertia principle aimed at the discharge of tension and the reestablishment of a minimal energic state is regulated by "a preformed apparatus that is strictly limited to the neuronic apparatus" (Lacan 1959–1960, 27). How could we not find here, in this concept of "preformed apparatus," an element of a neural or endocrine loop in a mechanism of physiological feedback? This suggests that in physiological functioning there are circuits destined to maintain homeostasis. This is clearly the case, as we have seen, for example, in neuroendocrine feedback loops. In mental terms, one could say that the pleasure principle makes use of preformed paths of drive discharge. What is more, these preformed circuits are themselves subject to adaptive mechanisms of plasticity. And in fact, we find this type of adaptive plasticity in the neuroendocrine systems (McEwen 1996).

In this connection Freudian theory introduces the concept of facilitation (1895, 1920),* which we can liken to the facilitation of information transfer established from experience by the mechanisms of neural plasticity. To return to Freudian terminology: the pleasure principle as the principle of inertia makes use of the facilitations that it preserves—indeed, consolidates—through pathways that are preformed within the neural apparatus. What is facilitated is the utilization of pathways of drive discharge so as to ensure that a certain level of energy will not be exceeded and that excitation will not become harmful to the person. These facilitations are preferential routes, a network of mutually correlative traces that make drive discharge easier. The process is self-maintaining; the more the drive is discharged through these pathways, the more the pathways are consolidated, promoting an

---

*Translator's note: Freud's term is *Bahnung*, translated into French as *frayage*; the literal meaning of both is "clearing a path." *Facilitation* is the customary English translation.

automatic type of functioning. Thus the pleasure principle is not only an inertia principle but also a repetition principle. In Freudian terms, facilitation is an established pathway, infinitely usable, through which the excess quantity of energy is destined to escape. In physiological terms, the reestablishment of homeostasis takes routes that are consolidated by the mechanisms of plasticity through repeated use.

If the pleasure principle is mediated by the discharge of endopsychic excitation, what becomes of this discharge? As we have seen (fig. 9.4 in chapter 9), the action resulting from the discharge of the drive will be perceived by the person himself as a stimulus coming from external reality. The situation is that a stimulus coming from internal reality is identified as coming from without on account of drive discharge. How does this occur? A person has an unconscious reality, inaccessible to perception. Through association with a somatic state, a drive is triggered and is discharged in accordance with the laws of homeostasis and hence of the pleasure principle. An action results from this discharge. The result of this action becomes a stimulus perceived by the person as coming from external reality, though it originated in unconscious internal reality (fig. 10.2). In this sense there is identity of perception[3] between what comes from unconscious internal reality and what comes from external reality. Ultimately, the person is dealing with a perception,

3. This also means that, according to Freud, there is identity of libidinal cathexes (Lacan 1959–1960, 31.)

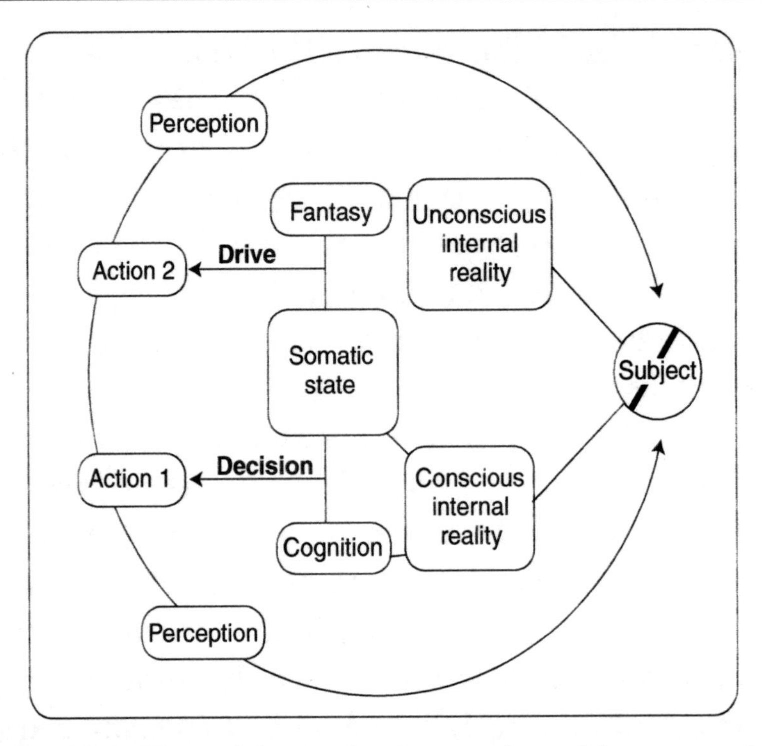

Figure 10.2 The subject, who is constituted by a conscious internal reality and an unconscious internal reality, is divided by the impact of the perception of actions emanating either from decision processes (Action 1) or drive discharge (Action 2).

either a real one coming from external reality, or what Freud calls a hallucinatory one coming from internal reality. Thus the primary process by which the unconscious functions (see chapter 9) affects perception indirectly by mediating the action resulting from the discharge of drive excitation in accordance with the pleasure principle. Through the process mobilized by drive discharge, the fantasy (that is by definition inaccessible to consciousness) reveals its presence via a perception of which the person is aware (fig. 10.2).

Thus the two destinies of the drive are combined. The one, of a physiological order, is to reestablish homeostasis; the other, of a mental order, is to signal to the person the existence of an unconscious internal reality that directs his actions even as it remains mysterious. The origin and destiny of the drive are bound together in a vital way, for the protective functions are as important as the functions of perception.[4] For we can be traumatized by our own endopsychic energy, as we see in therapy with very young children who have been left alone without the intervention of the other, the other being the sole means of discharging excitation. The entire organism, subject to the law of homeostasis (law of inertia) through the necessary discharge of excitation in accordance with the pleasure principle, thus seems made as much for protection against excitation as for perceiving it.

Let us now examine more closely the sequence mobilized in the identity of perception (fig. 10.2). A person's psychic life is constituted by an unconscious internal reality, which we have discussed in detail, and also, of course, by the conscious internal reality of the cognitive processes, conscious memories, and the result of various acts of learning. It is from these two elements that the person will organize his response and actions

4. "*Protection against* stimuli is an almost more important function for the living organism than *reception of* stimuli" (Freud 1920, 27).

when confronted by events (which we have also called external stimuli).

Let us look first at the response to this event in the case where the action (Action 1) is in direct relation with the stimulus (figs. 10.2 and 9.1). The person treats the information on the cognitive level, especially by associating it with awareness of the somatic state in which he will find himself after his decision. This is the very foundation of the theory of somatic markers (Damasio 1994). The anticipation of the somatic state will therefore determine in a basic way the decision leading to action (Action 1 in fig. 10.2). Thus we are in the register of the cognitive, colored by the anticipation of the somatic state, which leads to an action in direct relation to the event that has occurred. This action will be perceived by the person and will become for him a stimulus coming from external reality.

But an endopsychic excitation can arise from the interaction between the somatic state and fantasy, these being the constituents of unconscious internal reality. From this interaction emerges the drive that, in order to be discharged, leads to an action (fig. 10.2, Action 2). This action, in turn, becomes a stimulus perceived by the person as external. Hence we see that the person is dealing with two parallel perceptions. One was triggered by the occurrence of an event in the external world, leading to a response ruled by cognitive processes in relation to this stimulation. The other is interpreted as coming from the external world though it in fact originates as an endogenous stimulation. The person is thus divided by this identity of perception.

Let us take an example. A man gets up at night and, feeling very hungry, goes to open his refrigerator. He is not happy to do so, since he would rather go on a diet, but the tension is too strong and he does not resist. At this moment his wife joins him in the kitchen. He can't stand this, and, before she says anything, he rejects her, saying that she's always preventing him from doing what he wants.

What is going on here, of course, does not have to do with food. The calming effect that is supposed to come from the contents of the refrigerator involves another scenario that is, however, kept out of awareness. This man reproaches his wife for refusing herself to him, especially in their intimate life. Analytic work will, in fact, show him that the opposite is true. He is the one who is resisting, under the pressure of an unconscious scenario of the incestuous type. He discovers that, unconsciously, it is his mother that he desires through his wife, but his mother is forbidden to him. Hence his sexual inhibition, which he displaces onto his wife even as he reproaches her projectively[5] for refusing herself to him. In this

5. "[A] particular way is adopted of dealing with any internal excitations which produce too great an increase of unpleasure: there is a tendency to treat them as though they were acting, not from the inside, but from the outside, so that it may be possible to bring the shield against stimuli into operation as a means of defence against

woman he has chosen he is unconsciously seeking what attached him to his mother: her eyes, her scent, the way she acted toward him. "Eat, eat," his mother always used to tell him. "Clean your plate, grow up, become strong and handsome!" What we see here is a discontent bound up with incestuous desire. This is what prevents the man from feeling free with this woman he thinks he loves, even if it is another woman, his mother, that he is aiming at through her.

This tension is pervasive in him. He cannot find a way to discharge it except through a displaced object, though this object links up with his relation to his mother. It is the food that he eats to excess, though he should forgo it if he is to remain slim and seductive. He is caught in a vicious circle in which what is going on in his relation to external reality is in fact primarily determined by the pressure of an unconscious fantasy scenario of the incestuous type, a fantasy he has as yet not explored, the source of tension and discharge. There is confusion between the perception of the action connected to conscious reality (being hungry) and the perception determined by unconscious reality (incestuous desire). The perceptions of these actions converge toward a single action, opening the refrigerator, even though something entirely different is at the origin of this action.

---

them. This is the origin of *projection*, which is destined to play such a large part in the causation of pathological processes" (Freud 1920, 29; emphasis in original).

The identity of perception of the two actions (opening the refrigerator and attacking his wife) divides him. Hence his discontent. He does not know what has come over him. The attack on his wife is the effect of an internal conflict of which he is completely unaware. He does not recognize himself in this behavior. He thinks he is in conflict around food, but the impasse is located between wife and mother. His action toward this woman he loves troubles him. Why the aggressive attack? Is this really him? In this example of the man who is hungry, the identity of perception disturbs the sense of identity. The husband is divided between what is coming from the present situation and what is coming from an unconscious reality fortuitously activated by the pressure of a simple somatic state connected with hunger. The link between the two is the coinciding of the hunger he feels and the historical relation of his mother to food.

Luckily, we are not always affected by distress stemming from the identity of perception. A dialectic is established between what is determined by the pleasure principle via drive discharge and, on the other hand, the reality principle (fig. 10.3). For there is a contradictory relation between what comes from unconscious internal reality and the possibility of taking reality into account consciously, critically, and cognitively. The tendency to action stemming from drive discharge is subjected to reality testing. A person has the capacity for critical thinking. From the fact of being divided by the discontent within him, he can sense that he is getting further away from reality. He can thus correct the influence of a

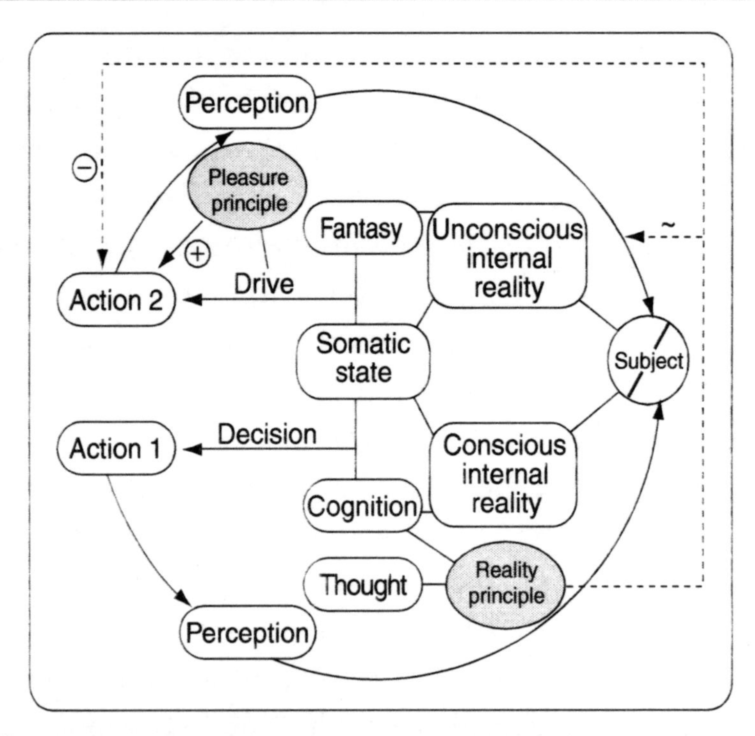

Figure 10.3 The reality principle modulates (~) the perception of actions produced by drive discharge, or it directly inhibits it (–). There is thus a dialectic between the unconsious pleasure principle and the reality principle linked to consciousness and thought.

perception of an action determined by the drive and return to the perception of the present situation. In our example, he can be troubled by his outburst of aggression, mistrust it, distance himself from it. The impulse can become problematic, not justified by the actual situation, and this puts him on the path to recognizing the unconscious incestuous desire involved in his behavior. From then on he can do something, reestablish a course of action consistent with reality, even if the discontent persists. This is how, for Freud, a reality principle joins the pleasure principle in a conflictual dialectic in which the former can take charge in a regulatory fashion that makes it possible to postpone a satisfaction coming directly from drive discharge (fig. 10.3).

The reality principle comes from what Freud calls the secondary process. In contrast to the primary process, it is based on contradiction; makes use of negation; acknowledges time and spatial differentiation; and, through the thought process formed by the activity of representation, allows for a suspension of discharge. As Freud writes, "Thinking was endowed with characteristics which made it possible for the mental apparatus to tolerate an increased tension of stimulus while the process of discharge was postponed" (1911, 221). For Lacan, however, the primary process tends toward the identity of perception and the secondary process tends toward an identity of thought.[6] What does this mean?

> It means that the interior functioning of the psychic apparatus . . . occurs as a kind of groping forward, a rectifying test, thanks to which the subject, led on by the discharges that follow along the *Bahnungen* already established, will conduct the series of tests or of detours that will gradually lead him to anastomosis and to moving beyond the testing of the surrounding system of different objects present at that moment of its experience (1959–1960, 31).

6. Lacan finds this idea in Chapter VII of *The Interpretation of Dreams* and also in the "Project." See Lacan 1959–1960, 31.

This leads Lacan to the paradoxical—or, at least, surprising—formulation that "[A]ll thought . . . occurs according to unconscious means" (1959–1960, 32). Such a statement is unexpected. At first glance, it seems that we must place thinking on the level of the reality principle, governed by the secondary process. Of course, as Lacan explains, it is not the pleasure principle that governs thinking. Nevertheless, thinking takes place in and from a field that is the field of the unconscious, and it is thereby subject to the pleasure principle.

Thinking makes possible the postponement of discharge. It is thus necessarily determined by the need for discharge and by the logic governing this need (which is precisely the logic of the unconscious, the primary process, and the pleasure principle), bringing into play representations having to do with unconscious wishes and pleasures that lie ahead. Hence all thinking really does occur in unconscious ways by virtue of the pressure of what is at stake in the unconscious world. This is its fundamental nature.

We have seen, then, how the fantasy scenario, in association with a somatic state, leads to a drive discharge in accordance with the pleasure principle. This involves a set of phenomena aimed at maintaining homeostasis in accordance with an inertia principle while respecting a principle of the constancy of variables comprising the internal milieu. All this is held by the connection between a somatic state and unconscious representations belonging to the fantasy scenario.

What happens if this connection could not be established? To answer this question, let us put ourselves in the hypothetical situation of the newborn and the state of distress in which it finds itself,[7] distress not yet resolved by the intervention of the other that would allow for discharge and for inscription in the form of an experiential trace.

Let us recall once again that, for Freud, the organism is not able to discharge itself alone, and that, subject to its own internal excitation, it is destined for a certain destruction of which distress is the sign. Like any organism, the infant is subject to all sorts of variations in its somatic state, variations that, for the most part, are reestablished in a physiological state with external support. For example, even temperature regulation is difficult at the beginning of life and requires care on the part of the other. The same is, of course, true of all the phenomena connected to feeding. Thus, as we have said, the functions of protection with regard to a living thing's excitation (somatic state) are of vital importance. The human child at the beginning of its life is subject to what might be called the *jouissance* of the living being, a crude, primitive *jouissance* not channeled by the other and, by the same token, not linked to representations. All that is present is the somatic, in the state of unpleasure we have described at such great length.

7. See Freud 1926, chapter 8 on the infant's state of distress (*Hilflosigkeit*).

Through the experience of satisfaction that requires the intervention of the other, traces are inscribed and in the process define trajectories of drive discharge. We might say that it is the other's specific action that gradually inscribes the trajectory of the drive. In any case, this is a special way of seeing the effect of education. The child is at first seized by somatic states, that is, by the force of the living being, not yet channeled by association with unconscious mental traces.[8] We know what a large role unconscious internal reality plays in the channeling of the living being's energy. In this sense, we might speak of a biological function of the unconscious for the survival of the individual. Without this association between somatic states and mental traces, the discharge of excitation could lead the living being to an inorganic state of complete disorganization without regard to the constancy principle necessary for survival. There would be an inertia that would totally flow away to the point of death, of return to an inanimate state, in a rapid exhaustion of life.

Let us go on to consider an individual we may suppose to be psychically organized, that is, someone who has been able to construct his unconscious internal reality by weaving the inscriptions and reinscriptions of traces associated with a somatic state. But let us also imagine that a disorganization occurs, unlinking the somatic states

8. These are, of course, also paralleled by traces inscribed on the conscious level through the systems of memory and the mechanisms of learning.

from the unconscious representations constituting the fantasy scenario. The elements constitutive of the drive's origin are disconnected, and the emergence of a drive obedient to the pleasure principle is no longer possible. This would be a situation well beyond the pleasure principle, one in which the *jouissance* of the living being wins out as in a state of primordial non-organization. The person finds himself invaded by somatic states of which he is the object, caught up in a process of desubjectivation, destruction, from the internal energy produced by somatic states unconnected with any representation. He becomes the object of the living being. This would entail the self-destruction of the person issuing from the somatic state. This is what Freud tried to convey in the notion of the death instinct, a term that, for him, refers to a fundamental category of drives that, by their systematic tendency to destruction, are opposed to the life drives (1920).

Let us sum up: somatic states that are not channeled by the possibility of a drive discharge resulting from their association with representations of the fantasy scenario lead to a destruction of the subject. We have seen how this could identify a fundamental biological function of unconscious internal reality. This raises the question of knowing why unchanneled somatic states lead to self-destructive behaviors.

One way of approaching this question is to say that biological systems and, in particular, the human being, have a spontaneous, preprogrammed mode of functioning that mobilizes the self-destructive processes Freud calls the death drive. This is, certainly, a risky *a priori* hypothesis, for we usually imagine that biological systems tend toward homeostasis and hence toward the maintenance of the organism's integrity. Yet this line of reasoning is the only way to understand behaviors that do not obey the pleasure principle, and, as Freud says, lie beyond it.

To make an analogy from thermodynamics, we might say that the organization of the living being may tend toward maximum entropy, toward disorganization and hence death. The very principle of the living being, in this view, would be death. By channeling this natural tendency in which the living being has the upper hand in a process of disorganization, the unconscious, and especially the fantasy scenario, make it possible to channel the entropy of the system and maintain it in a self-regulating, retroactive loop. Thus the living being has a spontaneous tendency toward death, dispersion, entropy, and hence self-destruction, and the unconscious is what makes it possible to organize somatic states into drives that enter into a physiological system aimed at the maintenance of the internal milieu and homeostasis.

11

Freud and James: Let's Be Synthetic

Every organism—and man is no exception to the rule—is from the physiological standpoint an entity responding to stimuli with motor acts.[1] In other words, to put it simply, external reality is perceived by the sensory systems, which trigger the appropriate motor response. This is, ultimately, the model of the reflex arc; it calls for no further elaboration.

But the discharge of the reflex arc is not all there is.

---

1. "The first thing that strikes us is that [the mental] apparatus, compounded of ψ-systems, has a sense or direction. All our psychical activity starts from stimuli (whether internal or external) and ends in innervations. Accordingly, we shall ascribe a sensory and a motor end to the apparatus. At the sensory end there lies a system which receives perceptions; at the motor end there lies another, which opens the gateway to motor activity. Psychical processes advance in general from the perceptual end to the motor end." (Freud 1920, 537)

External perceptions can also leave synaptic traces that are inscribed in the neural network by the mechanisms of plasticity. These synaptic traces are the neurobiological correlates of what, following Freud, we have called signs of perception. Yet this sequence of very simple events, which we could liken, for example, to the reflexive retraction of the siphon in Aplysia (Kandel 2001a), is insufficient to account for more complex aspects of human behavior.

As we saw in the preceding chapter, the theory of somatic markers suggests that a given perception is associated with a somatic state. This is the basis of the theory of the emotions and the mechanisms of decision-making that lead to action.

We can, therefore, add another dimension to the elementary model of the direct relation between perception and motor response: somatic markers make it possible to predict, on the basis of constructed representations, what the result of the motor response will be. Because of this, the perception–action reflex loop is strongly modified by the emotional aspect connected with the somatic state associated with a perception. Yet this concept remains in the domain of consciousness, for external reality emerges into consciousness by being inextricably linked to a somatic state.

And there is a third aspect: the unconscious and the constitution of a new internal reality unique to each individual. Whereas external physical and biological reality is the same for everyone, internal reality is inevitably singular, peculiar to each of us. Though perceived external reality is colored by a somatic state that can lead to a motor response, at the same time the perception of external reality fosters another reality, an internal one under permanent construction, that can modify the motor response, or, in turn, produce others (see chapter 9). It is in this sense that we can interpret Freud's statement that "all presentations issue from perceptions" (1925, 237).[2] As a result of successive transcriptions, the laws governing internal reality lead to a new complexity in which the signifier, which no longer has a simple relation to the signified of external reality, corresponds to a new signified, internal reality consisting of a chain of signifiers that obey laws other than that of material (physical and biological) reality.

The process of plasticity transforms the signifiers linked to external reality. Caught in a chain of associations, these move toward other signifiers that no longer

---

2. See also Freud's comments on the metaphor of the mystical writing pad, in which he discusses the infinite capacity of the psychic apparatus to receive new traces of afferent perceptions. This, he says, has to do with the fact that "unlimited receptive capacity and retention of permanent traces seem to be mutually exclusive properties," though he points out that "the inexplicable phenomenon of consciousness arises in the perceptual system *instead of* the permanent traces" (1925b, 227–228; emphasis in original). The idea that conscious perception and memory are mutually exclusive was already present in Freud's letter to Wilhelm Fliess of December 6, 1896 (Freud 1887–1902). In this context it is remarkable to note that Freud considers consciousness to be more inexplicable than the unconscious.

correspond to the signifieds of external reality. This is how internal reality is constituted. In other words, although, early on, words and letters—hence the signifier—correspond to the signified of external reality, that is, to an object or a situation, through a shift that occurs on the unconscious level the signifier is associated with a chain of other signifiers in such a way as to produce a new signified. This means that the same signifier can be associated with an external reality and, at the same time, with another signified in the internal reality under permanent construction, to the point where it is cut off from the original signified, which is entirely lost. In the psychoanalytic process the search for these original signifiers reveals the constitutive elements of the internal world.

Strangely enough, we find a point of convergence here between neurobiology and the concept of analytic treatment, insofar as a signifier is the equivalent not only of a sign of perception but also of a synaptic trace. This is one of the connections between the two incommensurables discussed in chapter 1.

Thus the characteristic of this internal reality that is gradually constructed by the mechanisms of synaptic plasticity over the course of experience is that it is organized in accordance with a logic in which signifiers become associated by escaping our conscious codification and constituting a restrictive fantasy scenario. An internal reality is formed in this way, one that has logical sequences, other than those of external reality, by which signifiers are organized in scenarios corresponding to fantasies. Jean Cocteau conveys a similar idea when, referring to history and mythology, he says that history is made up of truths that gradually become lies, whereas mythology is made up of lies that turn into the truth. In this image, history corresponds to the perception of external reality, which is transformed and reorganized in a fantasy that, we suggest, corresponds to the internal reality that gradually becomes a person's reality.

These new signifiers of the unconscious, organized in a fantasy scenario, are closely associated with somatic states and constitute the somatic anchoring of the drive. Beginning with the work of William James, the association between perception of external reality and somatic state has been the biological basis of emotion in the domain of consciousness. In a parallel analogy, we could say that the association between the signifiers of internal reality and a somatic state is the origin of the drive. The important point here, one that brings onstage the third aspect we mentioned above, is that this internal reality is perceived and integrated in the same way as the perception of external reality. In fact, we might say that the perception of external reality constitutes a sensory physiology, whereas the perception of internal reality constitutes a physiology of the unconscious.

The perceptions of internal reality are just as strong and striking as those of external reality. They provide fuel and are integrated on the same operational level so as to generate an action, a behavior. It is the distinctive function of psychoanalysis to decode this physiology of the unconscious by setting out from the signifiers that, in the unconscious, are associated with signifieds that no longer correspond to the signifieds of the code of external reality.

But at what level does the integration of the perceptions of external reality and those coming from internal reality take place? This will be discussed in chapter 13. According to the theory of somatic markers, a given somatic state makes it possible to guide the conscious decision-making process. This process aims at maintaining a homeostasis of the inner milieu, or in any case at avoiding an unpleasant somatic state. Similarly, it could be said that the perception of internal reality will influence the final decision taken to direct a given behavior, the goal being to reestablish or establish the lowest energic state. What is noteworthy is that we can place external reality parallel to the internal reality in which, each time, we find a link between perception and somatic state, opening out, on the conscious level, onto emotion, and, on the unconscious level, onto the drive. In both cases the result is an action for the purpose of maintaining homeostasis (see figs. 10.2 and 10.3 in chapter 10). The action has this aim when it is triggered by stimuli from either external or internal reality.

Thus we contend that the work of psychoanalysis is to decode internal reality by including the processes peculiar to somatic states, that is, by referring in a fundamental way to the drive dimension, so as to allow for direct access to external reality and make possible an action free of the fantasy constructions that so greatly interfered with it.

12

Redibis Non Morieris: *The Plasticity of Becoming and the Becoming of Plasticity*

A person's development is most often examined retrospectively, giving the impression that a history unfolds continuously in a series of causal concatenations. Yet the person's reality is different. At each moment it potentially remains subject to radical unexpectedness in its development.

This is especially true for mental development, which cannot be reduced either to the idea of a preprogrammed course or to that of direct mental causality. For we cannot simply connect a lived experience with its subjective effect without taking into account the ways in which the person may respond. On the psychic level, we may say that the combination of multiple determinants leads to effects that are not predictable *a priori.*

The same is true on the organic level through the phenomenon of plasticity, in which multiple epigenetic factors influencing the organization of the neural network beyond all genetic determination lead to a fundamentally unpredictable course of development. Experience is inscribed. It leaves a trace, and this trace is determinative. Thus determination is based on synchrony, that is, in the simultaneous instant of the event, by the reworking of the neural network corresponding to the establishment of a trace.

But what of the link between the traces, from one to the next, determinative for the individual's process of becoming: is it determined? What is determined on the synchronic level is perhaps not fully determined on the diachronic level, that is, in the successive concatenation of traces with one another. From one reworking to the next, the variability of responses increases, distancing the person from his determinants. In this way, the epigenetic process, whose plasticity is an operator, separates the person from his genetic determination. We see how greatly plasticity foregrounds the paradox of determinants, genetic and epigenetic, that leave the person open to changes that may be multiple and unpredictable.

In any case, there remains the whole question of how a material reality can be transformed into a mental reality and vice versa. In order to conceptualize this change we must understand the processes by which the identity of facts we consider to be connected are transmitted and maintained. We may well imagine that what happens is not a simple passage from one dimension to another. From the beginning Freud skeptically intuited that the chain of physiological processes in the nervous system is not causally related to the psychic processes (1891).[1]

Be that as it may, plasticity calls for a new approach to the issue of determinism. Through the structural and functional modifications produced by experience, it introduces the possibility of change. The process of becoming is neither determined nor undetermined: it is plastic.

To begin to understand the role of plasticity, especially the reworkings it can produce in the neural circuits both synchronically and diachronically, let us consider an extremely simple case (fig. 12.1).

Imagine stimulus S^1 perceived at time t^1 when the neural network is in a given state E. This stimulus S^1 treated by the neural network in state E will generate a response we shall call A^1. Through the mechanisms of plasticity, stimulus S^1 will leave a trace, inscription I^1 in the neural network, thereby reworking it. Thereafter the neural network will be in a state we shall call E^1 at time t^2.

If at this moment a second stimulus occurs, let us say, for simplicity's sake, that stimulus 2 is equal to stimulus 1, S^2 will produce a response we shall call A^2. Response A^2 may be identical to A^1, but it is conceivable that the reworkings that took place between E and E' makes response A^2 different from A^1. We see that plasticity has introduced a certain variability, hence also a certain degree of indetermination with regard to the same initial

1. Later, however, he was more optimistic about the contribution of biology to psychoanalysis; see the citation on p. xv.

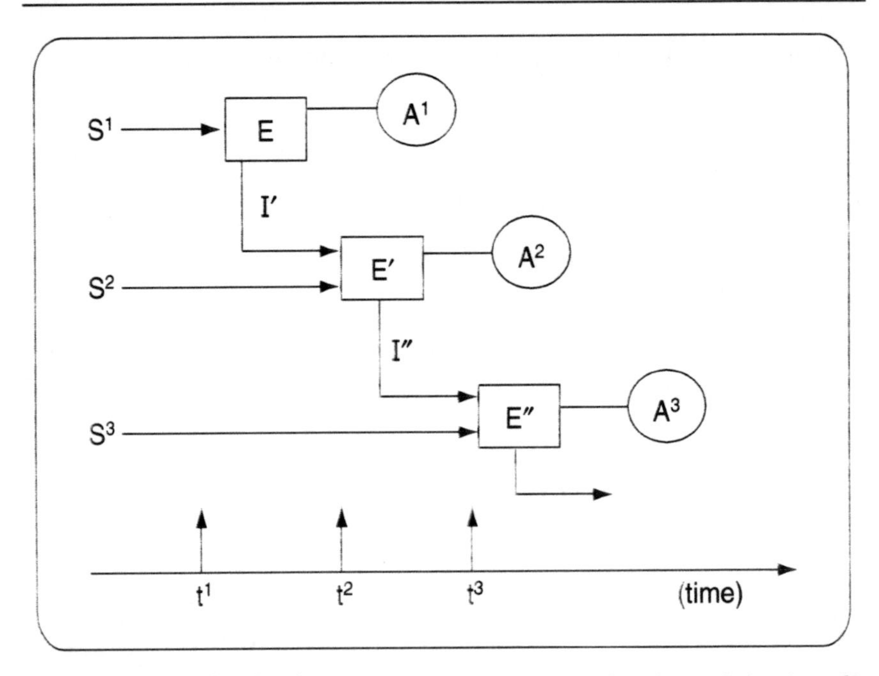

Figure 12.1 S^1, S^2, S^3 are three successive (diachronic) stimuli. E is the basal state of the neural network at the moment S^1 is perceived. E′ is the state of the neural network modified by inscription I' following stimulus S^1. E″ is the state of the neural network after inscription I″ following stimulus S^2. Similarly for S^3. A^1 is the action produced in response to S^1 when the neural network is in state E. A^2 is the action produced in response to S^2 when the neural network is in state E′. Similarly for A^3 and so forth. A^1, A^2, and A^3 may be the same or different; in the latter case, plasticity has introduced variability.

stimulus. For, in response to this stimulus, at time t^2 there may be a new response, different from A^1, which was not necessarily imaginable at t^1.

We can go further and say that, at time t^3, one and the same stimulus, S^3, will act on the system, which, this time, is in state E″. Response A^3 may be equal to A^1, or to A^2, or it may be different, which will introduce an additional degree of variability. In this intentionally simplified model, we have presumed that stimuli S^1, S^2, and S^3 are identical. If they were different, the variability of the system's responses would be even stronger.

The mechanism of plasticity in a temporal sequence of the diachronic process of becoming reworks the neural circuits in such a way that an identical stimulus can give variable responses. Plasticity thus introduces a degree of variability in the responses, which distances the neural network from the univocal, determined response we would find in a rigid system fixed in time. As the English neurobiologist Robert Turner has said, we never use the same brain twice.

As it introduces this degree of variability, plasticity acts within certain limits that are those of the neural network in place, and the variability of response introduced by the mechanisms of plasticity does not necessarily imply total freedom of the response. This is because the neural network and the very mechanisms of plasticity are subject to biological constraints that, though they open up a very large area of variability, are expressed in a context of biological determination. The very phenomenon of plasticity, in any case, requires us to think in terms of complexity resulting from a determination of variability, or, in other words, a determination of randomness.

It is not our intention to enter a debate on the concept of determinism from a philosophical perspective. Instead we are taking the side of physiology and simply saying that we are dealing with a system that, via the mechanisms of plasticity, is rearranged from one stimulus to the next, which results in the fact that in diachrony—that is, over time—one and the same stimulus can lead to different responses depending on the state of the system. In combination with plasticity, diachrony thus establishes an important degree of variability in a system that, in the absence of plasticity, would function deterministically.

This discussion entails a return to the notion of diachrony, which is at the center of the idea of becoming and causality. It would be absurd to hold that, from inscription to inscription, there is no continuity whatsoever. We have to accept the idea of a certain diachronic identity, for diachrony makes identity possible. We are obliged to posit that a certain identity persists over time, without which we would be absolutely unable to link up two states of the psyche, or two somatic states, or a somatic state and a psychic state. The absence of diachronic permanence would imply a loss of self-recognition on the part of the person and his connection to his own history, as we observe in the dementias or the confused dreamlike states of the psychoses.

Diachronic identity has been a central issue in the work of the linguist Ferdinand de Saussure in Geneva. The word *diachrony* is a neologism introduced by Saussure; he uses it, first of all, to describe an identity over time.[2] Freud, too, became attached, in his fashion,

2. Claudia Mejia, a colleague in the field of linguistics, has called our attention to an excerpt from Saussure's course of 1908, known through the notes of his students, in which he introduces the neologism. At every moment, he says, we face an identity over time,

to the diachronic connection, though he does not use this word explicitly. In "The Claims of Psycho-Analysis to Scientific Interest" he notes the diachronic nature of his discipline: "Not every analysis of psychological phenomena deserves the name of psycho-analysis. The latter implies more than the mere analysis of composite phenomena into simpler ones. It consists in tracing back one psychical structure to another which preceded it in time and out of which it developed" (1913, 182–183).

In this context we should distinguish conscious reality from unconscious phenomena. Conscious, cognitive reality develops sequentially, each element being in a relation of causality with the preceding one and the following one. What is learned synchronically is inscribed in a diachronic chain. We construct knowledge element by element; we learn addition and subtraction before differential equations.

Things are different in that unconscious internal reality that is the fantasy scenario. Here continuity and discontinuity, synchrony and diachrony, exist at the same time. One can surmise this, at least, from the theory of fantasy we have set out. If it is true that primary traces associate with each other to constitute unconscious internal reality, the diachronic dimension, represented by the association of these new traces over the course of

---

and we can even suggest a term, *diachronic*, meaning *across time.* Saussure goes on to reflect on how mysterious this diachronic identity is.

time, leads to constructions that are impossible to predict. Fantasy contains the diachronic dimension in a synchronic structure; time collapses, for determinism and unpredictability can coexist in fantasy. The diachrony of the unpredictable event is caught up in the synchrony of the structure of the fantasy. This is a result of the non-dimensional nature of fantasy.

We find the same phenomenon in the structure of language, which allows for ambiguity in signification. From the same phonemes we can arrive at opposite meanings. This is characteristic of the witticism, which is a special type of formation of the unconscious. We may cite as an example the double reading possible for the Latin phrase *Ibis redibis non morieris in bello.* The sequence of words is by nature diachronic, but a synchronic association can totally change the meaning. If we associate around *redibis non,* the meaning is that you have left, have not returned, and have died in war. If we associate around *non morieris,* the meaning is that you have left, have returned, and did not die in war. Thus we have two opposite scenarios produced by two different synchronic associations on the basis of a single diachronic sequence.

Let us continue. If we locate ourselves in a cognitive register, the diachronic concatenation of the traces is preserved. In cognitive learning, we construct knowledge that is sequential and successive. In fantasy, the traces are associated synchronically, not inscribed diachronically. Hence determinism (diachronic inscription of traces) and unpredictability (synchronic association) coexist. There is a determinism conveyed by inscription and an unpredictability produced by the possibility of synchronic association.

The distinctive task of psychoanalysis is to follow the thread of these synchronic associations. If this were not the case, psychoanalysis would be mere anamnesis, consisting in the restoration of a presumed diachronic identity. All that would have to be done is to grasp one's life as a chronological narrative. At each stage of a person's history, however, each stage of development, a synchronic juncture occurs, producing unpredictability and interfering with the story. To put it differently, what we find here is Freud's statement, cited above, namely that psychoanalysis is not just the decomposition into its constituent elements of what appears continuous; we need to take into account the synchronic network of associations constituting the very essence of the fantasy scenario. This is why fantasy is so insistent in a person's life: it makes its presence known in synchrony on the basis of all kinds of diachronic stimulations.

But we must highlight the role that may be played by the traces inscribed in the wake of a trauma. These traces, inscribed in diachrony, may be preferentially associated with other traces in synchrony. A network of traces interconnected by the trauma would tend to impose itself with particular force in the person's life, thereby introducing a deterministic tension into the unpredictable potentiality of synchrony. Like certain psychopathological disturbances, trauma could be seen as a disease of plasticity (Sandi 2003).

Fantasy is in itself a synchronic constraint that disturbs the mental handling of external reality, placing it in a context that is determined (fig. 9.1) and determinative. The person sees the world through the window of his fantasy (Lacan 2001). Thus we have a twofold determination: a diachronic determination connected with the inscription of the traces of experience and a synchronic determination connected with the associations produced by the fantasy scenario.

The structure of the fantasy partially determines the synchronic associations that interfere with the handling of external reality. Unconscious internal reality becomes determinative. The degrees of freedom of synchronic unpredictability are not infinite but are determined by the base material, the diachronic traces organized in a defined scenario.

We must, however, think of discontinuity in terms of a relation to a certain diachronic identity. We do not redo our lives each morning, even though we could potentially do so. The problem is knowing how a diachronic identity is constituted from mechanisms that would, in fact, allow for extreme variability. Here we encounter, in a certain way, the concepts set forth by Prigogine (1996), in particular, regarding the thermodynamics of irreversible processes; the new states of reworked matter determine what will happen to the extent that the processes involved in the reworking can be shown to be irreversible. But we also encounter the discussions of prevention and prediction in the domain of clinical psychoanalysis, where, despite the infinite possibilities of choice that a person has, he can be marked by his history or his fantasy in a restrictive, closed repetition.[3]

Perhaps we should use new terms in approaching the question of discontinuity and diachronic identity. Alongside *diachronic* we could have *becoming*, contrasting it with the development that remains largely caught in a deterministic perspective. The term *becoming* enables us to show the fortuitous nature of the result of a mental evolution. What we have to do is explain what occurred in order for a new state to be constituted by starting from a line of thinking that connects continuity and discontinuity. We may be saturated with determinants, but the connection between these determinants is not entirely determined. As Freud says, nothing is lost in mental life[4]; as present-day neuroscientists say, experience leaves a trace. The fact that all this is determined does not mean that it is predictable. Plasticity shows that everything is inscribed, that experience leaves a trace, and that this trace is determinative. Yet we are powerless to predict the process of becoming that it implies. Thus we cannot reduce this impossibility of grasping the process of becoming to the retrospective illusion characteristic of the

---

3. Freud speaks of the unconscious compulsion to repeat that must be posited to explain certain forms of resistance to change in analytic treatment (1920).

4. This is the view that "in mental life nothing which has once been formed can perish—that everything is somehow preserved and that in suitable circumstances . . . it can once more be brought to light" (1929, 69).

concept of development. This concept must, in any case, be modified by taking into account the element of unpredictability in a person's process of becoming, beyond both his mental and his organic determinants, both genetic and epigenetic. To think in terms of plasticity is to think in terms of a dialectic between these lines of determinants (Malabou 1996, 2000, 2004).

13

The Couple at a Red Light: The Influences of Internal Reality

The main argument we have been developing up to now is simple to summarize. Through the mechanisms of plasticity, experience inscribes traces in the neural circuits. Certain of these traces, which can be directly recalled to awareness, underlie memory and learning. Others can be rearranged, enter into association among themselves, and produce new traces that, in turn, are no longer in direct connection with the initial perception and can escape awareness. Finally, we might posit that certain perceptual traces are inscribed from the outset in systems that are inaccessible to awareness.

In this view, therefore, there are three kinds of traces: those that are directly conscious or able to be recalled to consciousness, those that escape consciousness secondarily through mechanisms of reassociation leading to a discontinuity between perception and trace, and those that are unconscious from the beginning.

For the conscious level, we can take the example of cognitive learning, motor learning, or even certain forms of emotional learning that make us consciously avoid situations we know are unpleasant or seek those that *are* pleasant. For example, we learn that not doing our homework involves punishment and doing it may lead to a reward. Such a view, however, does not take into account what clinical psychoanalysis teaches us, namely that a person does not necessarily want what is good for him,[1] and that, among several possibilities, in dealing with failure he will sometimes first aim at the one that leads to unpleasure. In so doing he is responding to strategies it is hard to explain solely in terms of identifiable conscious processes.

What, then, determines action? We cannot eliminate the idea that perception in the moment interacts with both conscious traces and unconscious ones to direct the action. Hence we must imagine the existence of a cerebral system in which the different bits of information—perceptions coming directly from an external stimulation, conscious and unconscious mnemic traces—are integrated so as to determine the action. From the functional point of view, we describe this integration as being

1. The person may also seek *jouissance* in self-destructive behaviors and repetitions that end in the same failure every day, which led Freud to posit a repetition compulsion he found within psychoanalytic treatment itself, in the form of resistances he called the negative therapeutic reaction: the further along a patient gets in his analysis, the worse he feels. It is from this kind of facts that Freud drew the obvious conclusion that there is a death drive. See Freud 1923.

carried out by what is called working memory, which mobilizes regions of the brain located in the prefrontal cortex (Fuster 2000b; Goldman-Rakic 1999).

Let us first go back to the mnemic traces and their association with somatic states. A neural system seems to play a central role in this association, namely the amygdala, a region in the human brain having the shape and size of an almond (hence its name) and located on the inner face of the temporal cortex (see fig. 6.1 in chapter 6). As we saw in chapter 6, this region is activated by different sensory systems. Moreover, it is in relation with the autonomic and endocrine systems that determine somatic states. Finally, it extends to the prefrontal cortex, whose role in working memory has been mentioned (Smith and Jonides 1999; Garcia and colleagues 1999). The amygdala is thus a crossroads that joins perception and its inscription with the triggering of somatic responses, even as it provides working memory with information via its connections with the prefrontal cortex.

This neural system ties together the amygdala, the autonomic and endocrine systems, and the prefrontal cortex. These may be the sites at which traces are inscribed or their association with somatic states comes about so as to constitute an internal reality that intervenes in our behavior, sometimes modifying or disturbing it. Current data in neurobiology tend to indicate that traces inscribed in the amygdala are those that are unconscious from the outset (Morris, J. S., Öhman, and Dolan 1998, 1999).

A partial view of mental life could be summed up in the fact that our action is a direct response to perceptions coming from external reality in a sort of reflex model. At a level of greater complexity the influence of memory systems would come into play. In this case information that can be recalled to awareness is preserved, inscribed by the mechanisms of synaptic plasticity. Taking only these two levels into account, we would find the individual always responding simply and directly to the stimulations of external reality in a logic appropriate to the presenting stimulus. We could also add the impact of conscious recollections having an emotional connotation of pleasure or unpleasure.

Yet the view in which the human being is solely determined by the influence of perceptions and learning is highly reductive insofar as it completely brushes aside the question of a mental life that, though it is derived and elaborated from external perceptions, can generate its own stimuli.

We cannot conceptualize everything on the basis of the external stimulus. There is not necessarily a direct connection between the external stimulus and the action that is triggered. We must also take into account the possibility of internal stimuli coming from a mental life constructed, transcribed, and rearranged from external stimuli. Up to now we have spoken of internal reality, mental reality, endopsychic perception, and the network of traces associating these with each other in connection with specific somatic states. But in fact we must try each time to grasp the unconscious internal reality on which this book centers.

Experience leaves unconscious traces associated among themselves and with somatic states. These unconscious traces constitute an unconscious internal reality that, in turn, can produce specific stimuli. These also enter into relation with conscious states, indeed, with immediate perceptions, by engaging the working memory, and they contribute to the realization of an action in the face of a given situation.

Thus we must account for this unconscious internal reality made up of traces to which are associated somatic states characteristic of them. The question is how and why this unconscious internal reality (or fantasy scenario) helps to modify our perceptions of reality and determine our behavior and actions, that is, our relation with our familial, social, cultural, and professional environment. In other words, this fantasy scenario can either remain inactive, simply mobilizable by conscious will (like a kind of film archive in which we have scenarios that we can freely review or replay), or it can be permanently active, serving as a restrictive frame for all perceived reality and determining on a fantasmatic basis the relation to external reality and the direction of our actions.

Let us take a very simple example that brings into play a perception and the recall of elements linked to our past experience. We are in front of a pedestrian crosswalk and the light turns red. We will stop. We recall that the red light means that we must not cross and must wait for the light to turn green. But there may be variations in this behavior; for example, if on the street to be crossed there is very little traffic, if there are no cars left or right, we may decide to cross even if the light is red, especially if we are in a hurry. A mechanism of protection is clearly at work in this case. In this situation we take into account the external stimulus (the red or green light) of an action to be performed (crossing the street) and a modification of this situation by our memories of this type of context. We will not, in fact, simply react reflexively, automatically, to red and green but will evaluate all the elements of the context inscribed in our memory systems, and these will influence our final decision about what action to take.

Today we know that this kind of contextual evaluation, which involves both immediate perceptions and memories that make it possible for us to evaluate a situation globally, is effected by the particular memory system known as working memory. Its name tells us that it enables us to "work on" different pieces of information coming from both external reality and the conscious memories inscribed in our memory systems (Fuster 2000a; Smith and Jonides 1999; Baddely 1998).

Another typical example of working memory is keeping in mind a number we looked up in the phone book so that we can dial it on a telephone. Working memory is by definition temporary. The elements remain active in it during the time we need in order to evaluate the situation and perform the action.

We can see how simple our life would be if working memory were fueled only by immediate perceptions and conscious contextual memories. Our actions would always be directly connected to the immediate perception. We would be in a logic of the sign: perception would be a sign of something we would respond to directly and appropriately as a function of a goal to be attained. But we know well that things do not happen that way, in that there are other elements, belonging to mental life (especially unconscious life) that interfere with the decision-making process.

Let us return to the example of the red light. The light is red, but there is no car in sight and you are in a hurry. Yet you do not cross. Why? Either you refer to your upbringing, which included a prohibition mobilizing guilt and the superego, or you are overcome by a feeling of unease whose origin you do not understand but that paralyzes your ability to act. You have not become aware that, on the other side of the street is an advertising poster showing a happy couple riding in a convertible. This stimulus unconsciously recalls a relationship that unexpectedly ended during a trip. You have the most unpleasant memory of this, a memory that keeps coming back in nostalgia for this lost relationship. At the moment you are not thinking about it, at least not consciously. All that happens is that you are standing there, immobilized at the curb, on a deserted street in front of a harmless red light.

The activation of mnemic traces associated with the somatic state of unpleasure in connection with a history of frustration interferes with decision making to the point of inhibiting action. This unconscious internal reality in fact contributes in a determinative way to decision making.

How, then, is this unconscious internal reality, this fantasy scenario, activated so that it becomes involved in working memory? Here is where the amygdala seems to play an important role (see fig. 6.1). The question then becomes: how are the traces inscribed in the amygdala activated by an external stimulus (McCaugh 2004)? As we have seen, sensory stimuli can directly activate the amygdala without passing through the primary sensory cortical areas (LeDoux 1996). Let us take, for example, a visual stimulus that activates the receptor cells of the retina, which then sends projections to the thalamus, the principal relay for all sensory afferents (fig. 13.1). From the thalamus other neurons extend toward the areas of the cerebral cortex that deal with specific sensory modalities, in our case the primary visual cortex.

Parallel to this neural pathway from the thalamus there are projections directly to the amygdala. The amygdala is itself connected to the autonomic and endocrine systems that control somatic responses (LeDoux 1996, 2003; see fig. 6.1).

Thus there is a sensory pathway dealing with information in detailed fashion in a mode we may consider conscious; this is the pathway that leads to the cerebral cortex (LeDoux 1994, 2003). Yet a second pathway, short-circuiting the cerebral cortex by going directly from the thalamus to the amygdala and thence to the autonomic and neuroendocrine centers, links external stimuli to somatic states in a way that remains unconscious.

Experiments conducted on animals, and, more recently, through functional imaging in man, confirm that

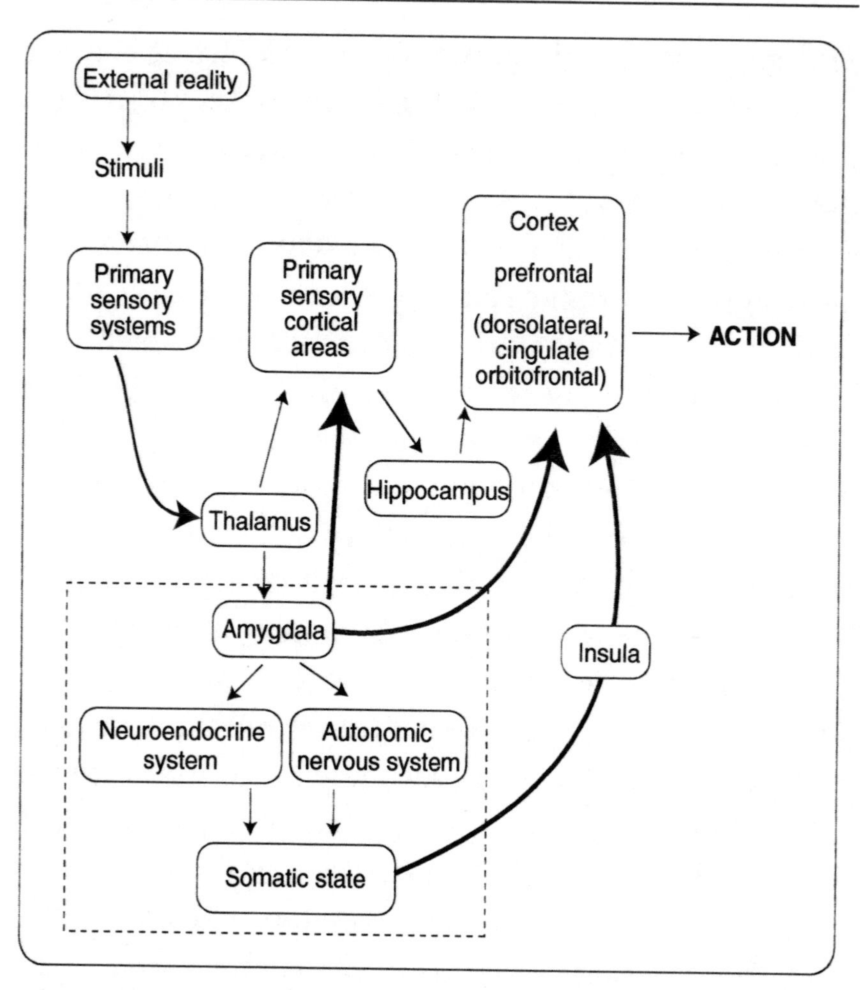

Figure 13.1 Neural circuits that can contribute to the constitution of unconscious internal reality and participate in the determination of action.

the amygdala plays a central role in joining sensory perception and somatic responses (LaBar et al. 1998). Moreover, synaptic facilitations in the form of long-term potentiation, can also form in the amygdala (Humeau et al. 2003). This is not to say that the unconscious is located in the amygdala but rather to suggest that the strategic position of the amygdala, at the interface between external stimuli and somatic responses, makes possible the reactivation of previously inscribed unconscious traces and the associated somatic states. This is one of the ways in which an external stimulus could activate a fantasy scenario and the somatic state connected with it, resulting in an acting out of the drive (see fig. 9.1 in chapter 9).

The drive requires a need for discharge through an object or an action in order to resolve the unpleasant, even distressing, somatic state into which the person has been plunged by the activation of his fantasy. The element of external reality that is able to activate the fantasy is most often an object or situation that is neutral in itself but possesses a high capacity for activating the fantasmatic aspects of unconscious internal reality and the associated somatic states peculiar to each subject. To return to the clinical field, let us take the example of objects that can activate sex drives leading to behavior at the far edge of societal rules, or even beyond that edge, to discharge the tension produced by these drives. *Psychopathia Sexualis*, the famous book by the Austrian psychiatrist Krafft-Ebing, provides a most complete catalog, based entirely on clinical cases, of objects and situations that can trigger irrepressible sexual drives (1886). This is especially true of objects invested in behaviors of the fetishistic type: a shoe, a foot, hair, or certain animals are all objects with a neutral valence for most people, but they can produce libidinal drives that, for some people, are so strong that they must sometimes be acted out.

Another example will help us understand some aspects of the neurobiological bases that can underlie such drive activations by objects or situations that are neutral in themselves. This concerns recent findings in the domain of addiction to psychoactive drugs like cocaine, heroin, or the amphetamines (Koob 2003). A familiar behavior in this domain is the reactivation of withdrawal symptoms[2] in people who seemed to have been freed of their addictions and had not compulsively sought the drug for a long time: several months or even several years (O'Brien 2003). They may manifest surprising withdrawal symptoms when they see objects as harmless as a phone booth, a piece of furniture, or a street corner (Childress et al. 1999).

For these inherently neutral situations to trigger withdrawal symptoms they must have been previously associated with drug use. For example, the phone booth recalls the one in which the addict called his dealer for a delivery. Similarly, the street corner or piece of furniture evokes the one on which the addict took the drug. Such examples may be multiplied to infinity, since the ability to become an activator of withdrawal symptoms is connected to each person's history. What should be noted here is that an inherently neutral stimulus activates an

2. These are pronounced somatic states, essentially neurovegetative in nature, such as increased heart rate, sweating, and intestinal cramps: a procession of somatic states partially explicable in terms of autonomic dysregulations.

entirely characteristic somatic state accompanied by objectively measurable perturbations of the neurovegetative system (sweating, palpitations, generalized malaise, etc.). These trigger what might legitimately be called a drive entailing an irresistible urge to be satisfied in action, in this case, the taking of a drug.

Now the regions of the brain that are activated when these former addicts are shown videos of situations or objects neutral in themselves, but linked to the experience of drug use for each of these individuals, have recently been made visible through brain imaging conducted with the help of positron-emission tomography. The sight of these objects or situations of high evocative valence triggered objectively measurable craving accompanied by activation of certain regions of the brain, precisely those that, as we have suggested, are involved in the activation of fantasmatic contents: the amygdala and the frontal regions like the anterior cingulate cortex. This heightened activity was selectively observed in the former addicts; in control subjects and former addicts who were shown videos of objects or situations not connected with drug use, these regions were not activated (Childress et al. 1999).

Activation of the fantasy scenario and the associated somatic state in former addicts is an illustrative example insofar as it sheds light, at least partially, on the neurobiological substrates of these reactivation behaviors. The reactivation of withdrawal symptoms is, clearly, an extreme situation, but it makes possible the precise experimental control of the behavioral context, the objects and situations linked to a somatic response.[3]

Does this mean that the amygdala circuits activated by an external stimulus can contribute to working memory and thereby to the mobilization of executive functions? In other words, can the mnemic traces and the associated somatic states brought into play when the amygdala is activated by a sensory system contribute on the same basis as external stimulation to the activation of working memory?

Many studies have identified the prefrontal cortex as the neuroanatomical substrate in which working memory operates (Fuster 2000; Goldman-Rakic 1999). Three subregions are chiefly involved: the dorsolateral prefrontal cortex; the medial prefrontal cortex or anterior cingulate cortex; and the ventral prefrontal cortex or orbitofrontal cortex. These three regions of the prefrontal cortex are closely interconnected, and studies based both on lesions and on brain imaging have shown that they come into play when working memory is activated (Smith and Jonides 1999). We may then posit the existence of connections between the amygdala and the prefrontal cortex in support of the hypothesis that

3. We might imagine (why not?) that this kind of study could be extended to other objects and situations that are able to activate fantasy scenarios and precise somatic states in psychoanalytic patients, to the extent that the activators emerge in the course of treatment in contexts definitely less colored by psychopathological connotations but certainly very meaningful for the patient.

the activation of the fantasy scenario and the associated somatic states contributes to the executive function of working memory.

This is indeed the case. The amygdala, especially the central amygdala, projects toward the anterior cingulate and orbitofrontal cortex, both of which are strongly connected to the other region of the prefrontal cortex, the dorsolateral prefrontal cortex (fig. 13.1; LeDoux 2003). In other words, the amygdala transfers information to working memory by its extensions to two of the three divisions of the prefrontal cortex. In addition, another indirect loop connects the amygdala to the prefrontal cortex, or to one of its subregions, the orbitofrontal cortex, which is chiefly involved in the detection of somatic states. This means that once the amygdala has been activated by a sensory stimulus, the prefrontal cortex receives information both directly from the amygdala and indirectly by the afferents carrying information from the internal organs about the somatic state (fig. 13.1).

Functional neuroanatomy, as it is understood today, thus supports the hypothesis that the fantasy scenario intervenes in the determination of action by providing information to the regions involved in working memory. It confirms the idea that the fantasy scenario and the associated somatic states are significant elements in the action undertaken by the individual.

What is more, the amygdala can actually modify working memory by pathways other than these direct ones through the projections connecting it to the orbitofrontal and cingulate cortex. It also extends to the primary sensory cortices (auditory or visual, for example), which provide working memory with basic information on the external situation. In other words, the amygdala not only provides information from the fantasy scenario directly to working memory, but it also modifies the perception of external reality at the earliest stages, thereby potentially influencing the nature of the information transmitted by the sensory relays to working memory (Weinberger 1998; McDonald 1998).

The view we have been setting forth gives a central role to the amygdala in the constitution of the fantasy scenario. Other regions are undoubtedly involved, and we would in no way want to adopt a reductionist or localizing approach, one that exclusively favors a localization of the fantasy scenario based on the establishment of synaptic traces in the amygdala. Yet it is obvious that the amygdala is a primary interface between the perception of external reality, the determination of somatic states, and the functioning of working memory and thus ultimately of taking action.

In addition, taking into account its essential importance in the emotional connotation given to stimuli coming from external reality, it is not impossible, as we see it, that the traces inscribed in the amygdala are the substrate, or one of the substrates, of the fantasy scenarios and the associated somatic states that constitute what we have called unconscious internal reality.

But the most important point is the decisive role of this unconscious internal reality. It is clear that a number of issues need to be dealt with, especially the unconscious nature of such internal reality. Perception of external reality, which leads via synaptic relays in the thalamus to the primary sensory cortices, can be consolidated in the form of a synaptic trace in the hippocampus, a region that seems to be strongly involved in explicit memory, that is, memory accessible to awareness (Bear et al. 2001). But perceptions by the somatosensory system can activate the basolateral amygdala directly via the initial thalamic relays and be consolidated in the form of traces that, at that moment, will remain unconscious. Cognitive neuroscientists today describe this type of memory as implicit—possibly, then, as unconscious memory. For our part, we think that this pathway, if it can play a partial role in the constitution of the fantasy scenario, is not the only one involved; traces initially inscribed in the networks of the amygdala can become associated with each other and reinscribed in such a way that they are no longer in relation with the external stimuli that produced them (see chapter 5). Seen in this light, the unconscious does not simply come about through the initial traces inscribed in the circuits of the amygdala. It comes primarily from the associations among the primary traces or signs of perception (see chapters 5 and 6), and as a result it cannot be limited to implicit memory.

Let us try to summarize where we are at this point. Unconscious internal reality and the associated somatic states interfere with action, which is determined at the prefrontal level by the processes of working memory and the executive functions. The amygdala, at the interface between sensory stimuli and the autonomic and endocrine systems controlling homeostasis, seems to contribute strongly to the constitution of this unconscious internal reality. We might then posit that there is a subtle equilibrium among the various stimuli activating working memory: perception of external reality, conscious mnemic traces, and unconscious fantasy scenarios. This equilibrium would be a physiological state. In pathological conditions, the fantasy scenario of unconscious internal reality would weigh in more heavily and would increasingly distance the person from information coming from external reality and the activation of conscious traces that are essentially cognitive in nature. Internal reality becomes a system of interference with the perception of external reality and its contextual elaboration.

But this is a pessimistic view. We can also say that unconscious internal reality modifies (to use a term with a less negative connotation) the perception of external reality, leading to a process of judgment and action that is highly individualized and unique to each person. If this internal reality did not exist, we would probably act in a very uniform manner, perhaps reflexively and automatically, for the cognitive and emotional memories attached to each person's experience would be unique just the same but certainly with less diversity and creativity. Unconscious internal reality is in fact what makes us unique beings.

14

The Hour of the Traces: The Unconscious, Memory, and Repression

If unconscious internal reality is constituted by an association among primary traces (see chapter 5), leading to new traces that remain unconscious; if the connection with the initial experiences gets lost throughout these processes of inscription, reinscription, and association; if there is no simple, direct link between the signified of the external world and the signified produced in unconscious internal reality, must we conclude that the unconscious cannot be put in direct relation to one or the other form of memory: explicit, implicit, procedural, or others as defined in the lexicon of neuropsychology (Bear, Connors, and Paradiso 2001)?

The traces constituting the fantasy scenario of unconscious internal reality are in fact very different from a given memory system. What we must do is speak properly of new mnemic traces peculiar to the unconscious and not necessarily reducible to one or the other structure dedicated to memory as defined and localized by the approach of cognitive neuropsychology. The unconscious, then, is not a memory but a system of rearranged mnemic traces that are not a reflection of the external reality that produced them. In that regard the irruptions of the unconscious are instead a disturbance of memory.

The unconscious is activated as a network of associations among traces characterized by specific somatic markers, inscribed in various brain structures that cannot actually be localized.

In this view, the mechanisms of synaptic plasticity have a twofold function: they provide mechanisms for a relatively faithful transcription of external reality at the same time as they open a path for the constitution of a newly created internal reality that is unique, peculiar to each person, and that itself becomes the source of stimuli and new perceptions. In that sense we can indeed say, *"To each his own brain,"* but also *"To each his unconscious internal reality."* Having access to it requires the work of analysis (taking this word in its etymological sense), which goes back along the thread of the interlinked signifiers constituting the unconscious internal reality that represents the person.[1] The unconscious must be defined dynamically, not in localizing terms.

1. See Lacan's (1960, 1964) statement that a signifier represents the subject for another signifier. (Translator's note: Lacanians use the term *subject* where psychoanalysts in the Anglo-American tradition say *person* or *individual*.)

The traces constituting this network of associations of the unconscious are no longer in direct connection with the initial experience of the external world. Direct access to the initial experience gets lost in the process of multiple reassociations. What we are dealing with instead is a complex scenario that formed in discontinuity with reality. Analytic work can lead to a reestablishment of this continuity by uncovering the fantasy scenario. This is what enables the patient to separate himself from it and regain a more direct relation to reality. It also offers him the possibility of inventing himself by integrating this unconscious dimension, making use of it, by going beyond its restrictions.

In this sense we can understand that a fantasy scenario does not become desensitized in a behavioral therapy. This latter form of treatment can be effective only if the relation between perception and inscription is preserved in a linear and univocal form; in other words, by acting on the primary traces defined in chapter 5. Such a treatment would involve somehow deconditioning facilitated synapses in relation to an event. In contrast, behavioral therapies cannot access the associations of secondary traces constituting the fantasy scenario. In order to reach those, we have to catch hold, as it were, of signifiers buried in the associative network constituting the fantasy scenario.

For example, in the case of an animal phobia not necessarily originating in a traumatic event involving interaction with an animal, it would be naive to think we could decondition the patient by showing him animals in a non-traumatic context so as to desensitize him when faced with this situation. Any object in reality no longer in relation to the animal in question may activate a network of associations somewhere in unconscious internal reality, bringing on the phobic manifestation. We are dealing not with a phobic object but rather with a phobic signifier acting as an all-purpose signifier.[2]

The fantasy scenario blurs all direct access to experience by the very process of its formation. It consists of traces and associations of traces that have grown apart from the initial experience, and the person can no longer enter into direct relation with what took place: the experience has gotten lost in the defiles of its inscription and forms an unconscious internal reality. And it is because it is the mental destiny of experience to get lost in the process of its inscription that the unconscious is not a kind of memory. It is not an explicit memory, because, by definition, the latter is not conscious. Nor is it a procedural memory. Though procedural memory is automatic, enabling us to carry out a whole series of actions without our necessarily being aware of what we are doing (as in driving a car), it nevertheless involves cognitive mechanisms that are fully able to be recalled to awareness and have nothing to do with the unconscious in

2. Consider, for example, the case of Little Hans (Freud 1909), where the horse phobia turned out to have several versions, from fear of falling from a horse to the fear of the blackness of their mouths, which went back to the enigma of the feminine. See also Lacan's (1956–1957) commentary on this case.

Freud's sense. We, for our part, refer to the unconscious —as an unconscious internal reality—in a way that includes an especially significant term: *reality*. For mechanisms of transcription and association create an internal reality that is distanced, indeed completely cut off, from the experience linked to external reality. Through this mechanism of distancing the trace gets lost, one might say, from the origin of the experience. The initial experience is buried, masked, but it remains highly active and able to influence the person's actions: there is indeed a reality, but an internal and unconscious one, in the meshes of which is hidden the experience of the external reality that can no longer be found.

The fantasy scenario is thus a rearrangement of traces that serve as the building blocks of unconscious internal reality. This implies that the unconscious is not a mirror of external reality. It leads to another logic than the one regulated by the event or lived experience. Unconscious internal reality produces its own stimuli that enter into play in the person's actions. If the unconscious is not a direct reproduction of external reality, it is also not merely the product of mechanisms distancing from awareness the elements of unconscious reality such as repression.[3] The unconscious is above all a rearrangement of traces in a fantasy scenario, these traces no longer

3. The overly exclusive view of the unconscious as produced by the repression of information unbearable by awareness seems to us to do away with the primary unconscious as a system of traces. What is repressed must first be inscribed in unconscious internal reality.

having a relation to the external experience that generated them.

This definition of unconscious internal reality identifies an initial form of the unconscious: a primary unconscious consisting of traces rearranged into building blocks. An image of this primary unconscious might be the Giacometti sculpture "The Hour of the Traces" that we mentioned in our preface. Giacometti said that in this work he represented elements of his unconscious arranged almost randomly, in any case not in such a way as to reflect any kind of external reality. We could have spoken, in this book, of the sculptures of the unconscious, which would be a new way of referring to works of plasticity, this time in direct reference to the plastic arts.

From a therapeutic perspective we can say that analytic treatment aims at seeking and grasping hold of the signifiers (signs of perception or synaptic traces; see chapter 5) that, in associating with other signifiers, produce a new unconscious trace and become one of the building blocks of unconscious internal reality. It is this primary signifier—in relation both to the external reality that produced it and to the signifying chain belonging to the fantasy scenario—that analytic work seeks to unmask. By identifying this signifier with its two aspects, located at the interface between internal and external reality, we gain access to the fantasy scenario in unconscious internal reality. This discovery, this unmasking, leads to the reassessment and remobilization of what had become crystalized, fixed in a restrictive scenario that repeatedly determined the person's action unbeknownst to him.[4] Identifying a signifier that is at the interface of the external reality from which it came and the internal reality of which it is a part does not necessarily mean finding a direct causality between an element of fantasy and external reality. Still, identifying this signifier enables the person to do something else with it in a burst of creativity, of new endeavor, counting on the ambiguity of signification so as to go beyond the internal construction in which this signifier had been caught up. Instead of being the victim of a causality brought about by the features of his life, by identifying this signifier the person can free himself from the constraints of the fantasy scenario and invent his own responses, for which he is, ultimately, responsible. The logic of psychoanalysis is thus more a logic of response than it is a logic of cause.

Nevertheless, though the fantasy scenario intrudes on awareness and disturbs the perception of reality and the conduct of action, certain elements of the fantasy scenario do not make it all the way to working memory and hence to awareness. They remain held within unconscious internal reality. In Freudian theory, this setting aside is due to the central mechanism of repression. But why are these movements coming from the fantasy scenario set apart from awareness by the mechanism of repression? The Freudian explanation, which is confirmed from the neurobiological perspective (see below),

4. The German term *unbewusst* is a clearer reference than *unconscious* to what is not known (*wissen*, "to know").

is that in fact the emergence into consciousness of these drive movements coming from the fantasy scenario causes the person an unbearable amount of unpleasure. Hence the avoidance of unpleasure is the motive and aim of repression (Freud 1915c).

Since the drive is a frontier concept between the somatic and the psychic, that is, it consists of an element from the fantasy scenario and one from the associated somatic state (see fig. 9.1 in chapter 9), the question that arises is whether repression operates globally on the set of both elements or separately on each of them. Freud's reply is that repression acts separately on the representative of the drive (the element from the fantasy scenario) and on what he calls the quantum of affect (the somatic state associated with the drive).[5]

In the light of recent findings in neurobiology, it is possible nowadays to discuss these two components of the drive once again: the element from the fantasy scenario and the associated somatic state. For Freud, the fate of the representative of the drive (which, in our terminology, corresponds to an element of the unconscious fantasy scenario) can hardly be anything other than disappearance from consciousness when it had previously been conscious or being held apart from consciousness when it is on the point of becoming conscious. Recent

5. "From this point on, in describing a case of repression, we shall have to follow up separately what, as a result of repression, becomes of the *idea*, and what becomes of the instinctual energy linked to it" (1915, 152).

studies show how neurobiology contributes to our understanding of the suppression of undesirable memories (Anderson and Green 2001; Anderson et al. 2004). It seems possible to identify specific neural mechanisms involved in the removal from awareness of recollections calling on declarative memory. Imaging studies show that this suppression—which, in Freudian terms, we might call the repression of undesired memories—involves an activation of the dorsolateral prefrontal regions and a decrease in the activity of the hippocampus. In other words, there seems to be a mechanism of active control via specific neural circuits for suppressing (in Freudian terms, repressing) unwanted memories. We can imagine that this mechanism is mobilized in order to repress elements of the fantasy scenario associated with the drive.[6]

Still, such a model concerns only cognitive processes operated on by conscious will. Thus we are very far from what repression is, since repression has to do with unconscious mental processes. It is difficult to imagine that conscious will could intervene on the level of the fantasy scenario, actively preventing the accession to awareness

6. An interesting aspect of the work of Anderson and colleagues (2001, 2004) is that the repetition of the mechanism of rejecting from consciousness unwanted memories has a negative effect on the very retention of these memories. Once the mechanisms of repression have been brought into play, later recall of these unpleasant memories becomes harder still. In fact, the degree to which these unpleasant memories are forgotten increases with the number of times access to awareness was prevented by the mechanism of repression.

of certain of these elements. Yet it is possible that mechanisms analogous to those we find in the voluntary, conscious suppression of unwanted memories are also at work in unconscious repression (Levy and Anderson 2002). Our critique is not intended to diminish the relevance of these experiments targeting cognitive, conscious processes. What we want to do is note that repression of aspects of the unconscious fantasy scenario occurs unbeknownst to the individual, and also that the repressed elements are associated with significant somatic states that can also interfere with the exercise of conscious will.

Let us return to the question of what becomes of the other aspect of the drive, the one that Freud (1915) calls the quantitative factor of the drive representative or the quantum of affect (the somatic state associated with the drive). According to Freud, there are three possibilities: the drive is repressed, in such a way that no trace of it can be found; or it appears in the form of an affect with a particular quantitative coloration; or it is transformed into an affect that the person clearly recognizes as anxiety.

Although one might possibly admit that the neurobiological mechanism involved in conscious, voluntary forgetting is at work in the unconscious repression of the representative of the drive (an element of the fantasy scenario), it is less likely that this mechanism is also at work in the repression of the affect of the somatic state (for Freud, the quantum of affect). Moreover, for Freud himself the repression of the representative of the drive can take place in situations in which the quantum of affect is not repressed.

Continuing with this idea, we may say that the repression of the somatic state associated with the fantasy element is in fact clearly less effective. A disturbed, unpleasant somatic state would thus persist even if the repression of the drive representative were at work and created what from the emotional point of view would be described as a state of anxiety, sadness, or frustration. The fact that repression of the somatic state is much less effective, more subject to failure, than repression of the associated fantasy element increases our understanding of why states of anxiety are so very often met with in clinical treatment.

Let us take an example Freud (1915) used to show the dissociation between the more or less successful repression of the drive representative and the fact that repression is not at work on the quantum of affect. This was an animal phobia the analysis of which showed that the drive movement was a libidinal attitude toward the father. Repression erased it from awareness, and the father no longer appeared as the object of libido, but a substitute was found in the form of an animal that could plausibly serve as an object of anxiety. According to Freud, this substitution occurred through displacement, following determinative connections in a particular way. Yet the quantitative element did not disappear; it was transformed into anxiety. The result was anxiety with regard to wolves in place of a demand for love addressed to the father. This was a fundamental failure of repression. Even if repression of the representation linked to the unconscious fantasy scenario did indeed occur, the associated quantum of affect had another fate. It was not really repressed but returned in the form of anxiety in accordance with the central law identified by Freud: what is repressed returns from the unconscious, disturbing the person's conscious life in an important way (fig. 14.1).

Let us go back to the unconscious internal reality constituted by initially conscious traces that, in the play of associations and retranscriptions, create new unconscious traces. It is entirely possible to imagine that certain primary traces coming from perception of the external world will be directly inscribed in an unconscious manner, without passing through secondary associations and transcriptions. We may think here of traces that can be inscribed in the circuits of the amygdala through primary sensory pathways extending directly from the thalamus to the amygdala (see fig 13.1 in chapter 13). We know that these pathways underlie perceptions that remain unconscious (LeDoux 1996). It seems reasonable to imagine that these perceptions that occur unconsciously can also leave synaptic traces that, without being subjected to the rearrangements and associations we have looked at earlier, promote an unconscious fantasy scenario from the outset and are the building blocks of this unconscious internal reality.

When all is said and done, this synopsis of the building blocks of unconscious internal reality seems to indicate that these blocks produce permanent endopsychic stimulations with no logical connection to external real-

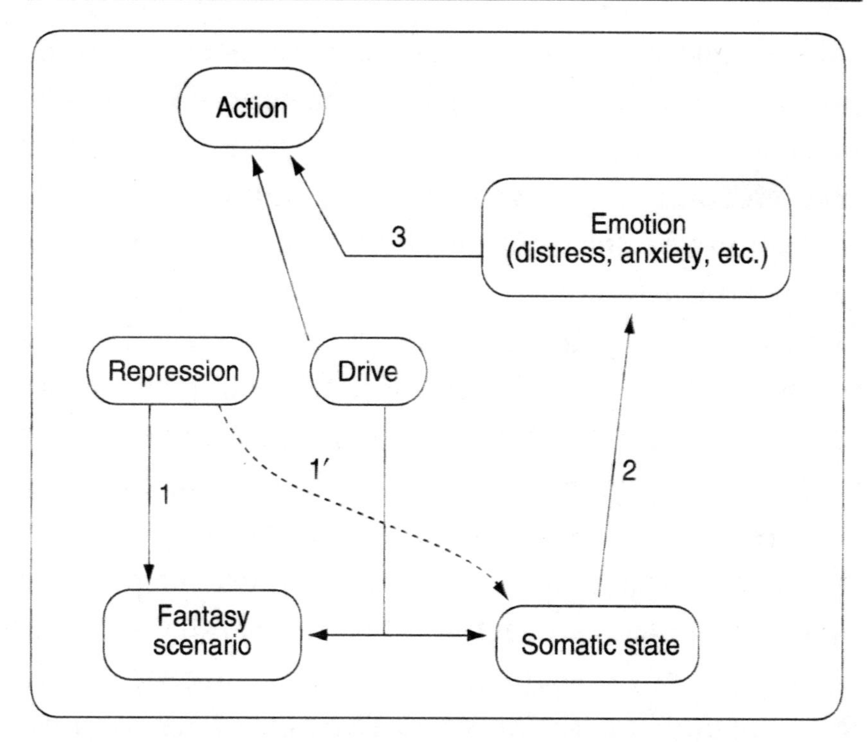

Figure 14.1 Repression, fantasy scenario, and somatic state.
1. Repression acts with relative success on the elements of the fantasy scenario (drive representative for Freud).
1′. Failure of repression of the somatic state (quantum of affect).
2. This somatic state (quantum of affect) is recognized as an unpleasant emotion, for example in the form of anxiety.
3. The emotional state disturbs the execution of the action.
If repression does not operate on these fantasy elements and their associated somatic states, the drive movement permanently intervenes in the person's actions (see fig. 9.1).

ity. This unconscious internal reality interferes with the person's executive functions, or, via repression, reactivates somatic states (such as distress and anxiety) that are perceived as unpleasant and difficult to manage. As we explore the constitution of this unconscious internal reality and its relations with a person's symptoms and behavior, we once again see its clinical impact. The disturbance of executive functions that it brings about can lead to actions that are inappropriate and create all sorts of problems for the person. The effect of the failure of repression on the somatic state gives rise to a whole series of unpleasant emotions that also disturb the person's behavior, choices, and actions. In any case, all these phenomena show the major role played by unconscious internal reality in a person's life. The neurosciences must account for this and, by the same token, become open to a new, necessary connection to what characterizes psychoanalytic theory and practice.

15

The Ferrari and the Trailer: Beyond the Fantasy Scenario

What determines the associations among the primary traces so as to generate these new traces constituting unconscious internal reality? This process remains mysterious. We might hypothesize that it is random, that by sheer chance two primary traces become associated and produce a new trace. We might also imagine a dynamic that orients the associations of traces, leading to the formation of a fantasy scenario. And does the fantasy scenario have a function? Does it correspond to a mental, or indeed a neurobiological, need?

Fantasy is a solution created to deal with a complex situation that poses many questions. It enables us to conceptualize the impossible, to integrate what Lacan described as the order of the real.[1] In the face of the real,

1. Lacan defines the real as "the domain of what subsists outside of symbolization" (1954, 388). Thus he defines the subject as a response to the real (1972). See also Miller 1987.

the child constructs all sorts of fictions. Hence the very early sexual theories existing in the unconscious: the child imagines himself impregnated through the ear and born through the navel or some other orifice. The child has countless sexual theories of this kind (Freud 1908, Ansermet 1999). As a little scientist (Freud 1910), he finds a solution in fantasy, thinking in terms of his surroundings, of his body, of what is mysterious for him.

In a case taken from clinical practice we find the following constellation: the father, the father's brother, a child of four or five (the patient), and his little brother. The father is the second-born sibling; he has been very unhappy about his older brother's excellence and feels somehow crushed by him. The message he sends his older son, the order he imparts, is that the boy must not crush the little brother who has just arrived, as he himself, the father, has been crushed. The critical point is that all the members of this family have succeeded at a high level professionally or socially. Thus the little boy is faced with an irreconcilable contradiction: he must be as outstanding as his elders but without overshadowing his little brother. How can he get out of this impasse?

Here is where a fantasy emerges, created from elements of external reality rearranged in accordance with a logic much subtler than one would imagine *a priori* and leading to all sorts of associations based on perceptions of the external world (at least, those the child has available to him at this time). It is a fantasy of a loser, a child who fails at everything he tries and will never be up to the challenges he undertakes to meet. The boy is constantly overcome by a feeling of dissatisfaction, a feeling that neutralizes the successes he achieves nonetheless. For this child does very well in school and is good at sports. In line with his family's tradition, he meets the challenges that face him. But his perception of his successes is colored by, or filtered through, a sense of failure and worthlessness, and this creates enough ambiguity to allow the child to obey his father's order even as he continues in his family's tradition of excellence.

We see here that fantasy can act as a solution to a problem that is impossible to manage otherwise. The child has found several relatively limited elements of his external reality to help him deal with an untenable injunction. These elements probably oriented the network of associations in his fantasy, which is not the result of chance but obeys its own dynamic. The fantasy enables the child to make pleasure and unpleasure coexist: pleasure linked to success but tinged with displeasure through the loser fantasy in which he sees himself as incompetent, even if this is not at all compatible with reality. This example also shows the extent to which fantasy is detached from reality.

This case involves a sort of *Ersatz* of unpleasure, relatively harmless insofar as it does not interfere too much with the pleasure of success, but, in a kind of deception, offers the advantage of making it possible for the child to obey his father's message. In using the terms *pleasure* and *unpleasure* we are of course referring to the somatic states associated with these conscious and unconscious movements. A pleasant somatic state will be associated with success, and a somatic state linked to unpleasure will be associated with the self-perception of incompetence. In this way, the pleasure associated with success is tempered, and the child can obey his father's order not to crush the little brother. This concatenation leads the child to see himself as a Ferrari towing a heavy trailer.

In this example the fantasy conceals and reveals. It is simultaneously an impasse and a solution. It forces reality into a frame that obliges the child to spend his time looking at the world through the window of his fantasy. It makes him interpret every success in terms affectively marked by the idea of a failure insidiously present or potentially on the horizon. This unpleasure represents a sort of tax that must be paid at every success. But this portion that has to be paid in terms of unpleasure is functional at the same time, since it enables the child to move ahead positively in his life and meet his challenges successfully.

In a general way the analysis of the contents of the fantasy helps us understand the principles guiding the associations of the primary traces among themselves so as to constitute unconscious internal reality. As we have seen, this is not a random process but one that corresponds to an inner economy, since it makes it possible to preserve pleasure in spite of everything, even if this means associating it with an unpleasure that is bearable and does not interfere with the person's actions and goals. A fantasy organized in this way is functional and can be seen as a solution to an impossible problem, a solution that emerges from the young child's confrontation with a real that he cannot deal with subjectively.

This example shows how fantasy is both an inscribed and constraining scenario and a solution, probably the least bad one, a ruse for getting out of an unthinkable impasse. We also see that it is the pleasure/unpleasure principle and the associated somatic states that regulate certain associations among primary traces in the form of a network of secondary traces belonging to the organization of the fantasy as a way out of an impasse. This fantasy solution is, to an extent, successful and well balanced relative to the constraints whose potential destructiveness might have come to the fore for the patient.

In other situations, however, the fantasy does not serve as a solution of this kind, as we see in an example discussed by Freud in 1919 in connection with a series of cases involving beating fantasies. Freud notes several stages in this fantasy. In the first of these, the child sees the father beat another child whom he, the child, hates. This is a lived experience, perceived consciously and attached to a satisfaction of the sadistic, or at least aggressive, type. A second phase, unconscious and reconstructed in analysis, presents a different scenario: the person administering the beating, that is, the father, remains the same, but the child being beaten is now the person himself. This version clearly has a masochistic quality that Freud interprets here as sadism turned against the self (see also Freud 1924). For in 1919 Freud had not yet developed the concept of primary masochism that he introduced with the hypothesis of a death drive, that is, of a pleasure that may also be found in unpleasure and tip over beyond the pleasure principle (as in the title of his 1920 study).

From this far side of the pleasure principle Lacan derives the notion of *jouissance,* the antinomy of pleasure. *Jouissance* carries the subject along in the flood of the drive movements that pervade him. He becomes the object of *jouissance,* including a self-destructive *jouissance.* In the case of the little boy, the fantasy had a certain functionality and the patient remained within the logic of pleasure and unpleasure. In the case of the fantasy in which a child is being beaten, the second phase, in which the person himself is being beaten by his father, can unconsciously determine behaviors aimed at getting caught up in a self-destructive *jouissance.* The drive movement comes to the fore, carrying the person along with it. He becomes its object and it determines his actions, though it does so without his awareness.[2]

Let us stay with this second phase of the fantasy, namely: "I am being beaten by Father." Although it remains unconscious, and even if it has never been true in reality, it is at work. In accordance with our model, we can say that associations are produced from primary traces (the image of the father administering the beating, the image of the other child being beaten, and other contextual elements) and lead to a series of secondary traces that become organized into a relatively simple fan-

2. In the terms of Figure 9.1 in chapter 9, Action 2 predominates, beyond the conflictual dialectic between Action 1 and Action 1′.

tasy scenario in which the person is himself the object of paternal aggression in a configuration that no longer corresponds to external reality. An unconscious internal reality linked to marked somatic states is constituted in this way, one in which the person has become the victim and experiences *jouissance.*

Freud goes on to describe a third phase, somewhat similar to the first: "A child is being beaten." This third phase, Freud notes, is especially frequent in certain analyses of hysterics and obsessional neuroses. It sometimes even reaches a climax in onanistic satisfaction, first with the person's consent but also later with a compulsive quality and against the person's will; the person has become the object of his own *jouissance.* Desubjectivation has taken place; the subject has become an object. The person doing the beating is no longer the father, and the person being beaten has become indeterminate, undifferentiated. It is no longer the other child hated by the person, nor is it the person himself, but "a" child, any child, without an identity. Only this last version of the scenario reaches consciousness.

This example illustrates a use of fantasy that is not a solution. The underlying unconscious fantasy, accompanied by very particular satisfactions and hence by a specific somatic state, is that of being beaten by one's father. It was produced by the experience of seeing the father beat a hated child. But it returns to awareness in a vague, although constraining, form, namely the obsessive idea of seeing a child being beaten by someone unknown so as to get and maintain an unconscious satisfaction. In this case the fantasy scenario is organized by secondary traces produced from primary traces. It goes in the direction of a subjective economy that does not work, that is constraining and ends in the expression of a perverse trait. This fantasy can thus be seen as a restrictive solution. In the end, it is really the scar of what remains at an impasse for the person, the aftereffect of a process that has ended.[3]

Let us go back to the very beginning. A child has to confront the enigma that he presents for himself. In his quest, he touches upon certain moments of nonmeaning. His place in relation to what has come before, his arrival in the world, his sex, the workings of his body: all this is at the same time familiar and strange to him. In order to manage the real that has come into play, he makes use of images and fictions produced by the associations among primary traces that come from external reality but are organized into a fantasy scenario corresponding to unconscious internal reality. This fantasy scenario certainly has many inconvenient features, if only because it is constraining, but it can also be said to be the sole possible way of trying to solve the enigmas or contradictions with which the child is confronted.

Even in more positive situations fantasy remains a solution. There is a certain price to be paid by the person, but one that on the whole enables him to live with the negative products of the fantasy (for example, neurotic

---

3. As Freud says, the beating fantasy "and other analogous perverse fixations would also only be precipitates of the Oedipus complex, scars, so to say, left behind after the process has ended" (1919, 193).

attitudes). In other cases, however, the fantasy becomes an excessive constraint, resulting in behaviors that make social interaction problematical and become a weight that is hard to bear. The solution is then dysfunctional.

In all these cases the work of analysis is aimed at making the person conscious of the fantasmatic nature of the scenario that he has constructed and that makes him see reality through a small window. It seeks to free the person from fantasy as the sole solution. Through a newly won freedom from the constraints of fantasy, the course of a psychoanalysis should enable the person to approach reality from a different perspective, to pass from the restrictions of an unconscious internal reality to the possibilities offered by whatever may happen.

The fact that the inscription of experience by the mechanisms of plasticity creates a distance from experience paradoxically offers a person freedom. It is what gives him room to move around, an ability to transform himself, to change, to become the author and actor of a process of becoming different from what was programmed by his determinants. Neural plasticity is thus a condition of a possible plasticity of becoming. Plasticity, finally, is what makes it possible for a person in analysis to free himself from the constraints of a rigid fantasy scenario or to make different use of the way it functions as a solution, to use the fantasy instead of being used by it.[4]

4. This statement calls for fuller discussion of the termination of an analysis, a complex issue, but our goal here is to look at what makes that termination possible. On this question see Freud 1937.

In short, we can define the psychoanalyst in a new way as a practitioner of plasticity, that is, someone who is counting on the potentialities of plasticity to reopen the field of possibilities, not by rejecting what came before, but on the contrary by using what came before to enable the patient to do something else with it.

Afterword

It comes as a surprise, at the end of this reexamination of Freudian theories, that concepts that might at first appear to be abstract elements of a theoretical construction effortlessly find a concrete physiological basis in the fact of neural plasticity. The trace, which is at the center of the phenomenon of plasticity, lies at the intersection of the neurosciences and psychoanalysis. What we have done in this book is show how the synaptic trace is related to the psychic trace and the signifier.[1] Through the association of somatic states, and traces left by experience, the psychoanalytic concepts of the unconscious and drives turn out to have a biological resonance. These concepts are fundamental in

1. In this connection see also the remarkable work of Jacques-Alain Miller (2000), who approaches biology from the field of psychoanalysis, especially with regard to the trace language leaves on the body.

both domains—psychoanalysis and the neurosciences —that at first seem to have nothing in common.

A question arises regarding the biological status of the unconscious and the drive. Why the unconscious? Why the drive? What is their biological function? These may not be the first questions confronting psychoanalysis. From the clinical perspective we first approach the unconscious and the drive in order to untangle what has been constituted in the form of an impasse at the whim of the person's life. In this context, the unconscious and the drive may remain useful concepts without the need for further attention to the question of their existence or their biological function. It might even be said that this question is ultimately of no importance in clinical psychoanalysis, which can work with complete effectiveness if it is never raised. On the other hand, as soon as we ascribe a biological status to the unconscious and the drive, the question of how they function returns to the foreground, calling upon neurobiology and psychoanalysis to work together and develop a mutual critique.[2]

A biology of the unconscious and the drive has been made possible today by the recent advances in the neurosciences that we have referred to in this book. We hope to have carried them further, offering psychoanalysis the verifications that Freud awaited from biology decades ago and offering the neurosciences a new access to specific

---

2. We are in complete agreement with Eric Kandel's (1998, 1999) proposals describing a new conceptual framework for the relations between biology and psychoanalysis.

questions in the field of exploration opened up by the hypothesis of the unconscious. What is at issue is not merely a logic of proof, demonstrating psychoanalysis on the basis of the neurosciences or persuading neuroscientists to take into account the theses of psychoanalysis, but rather seeing what follows, for each side, from the paradigm shift involved in the fact of plasticity, in which we find the extraordinary possibilities presented by contingent experience both for the process of becoming of each person and for each brain.

References

Aggleton, J. P., ed. 2000. *The Amygdala. A Functional Analysis.* New York: Oxford University Press.

Anderson, M. C. and Green, C. 2001. Suppressing unwanted memories by executive control. *Nature* 410:366–69.

——— Ochser, K. N., Kuhl, B., et al. 2004. Neural systems underlying the suppression of unwanted memories. *Science* 303:232–35.

Ansermet, F. 1999. *Clinique de l'origine. L'enfant entre la médicine et la psychanalyse.* Lausanne: Payot.

Atlan, H. 1999. *La fin du "tout génétique"? Vers de nouveaux paradigmes en biologie.* Paris: INRA.

Baddely, A. 1998. Working memory. *C. R. Acad. Sci. Paris, Sciences de la Vie, Life Sciences* 321:167–73.

Bear, M. F. 2003. Bidirectional synaptic plasticity: From theory to reality. *Philosophical Transactions of the Royal Society of London, B* 358:649–55.

——— Connors, B. W., and Paradiso, M. A. 2001. *Neuroscience. Exploring the Brain,* Second edition. Baltimore: Lippincott Williams & Wilkins.

Bernard, C. 1865. *Introduction à l'étude de la médecine expérimentale.* Paris: Delgrave, 1989.

Blake, D. T., Byll, N. N., and Merzenich, M. 2002. Representation of the hand in the cerebral cortex. *Behavioral Brain Research* 135:179–84.

Bliss, T. V., Collingridge, G. L., and Morris, R. G. M. 2003. Long-term potentiation: enhancing neuroscience for 30 years. *Philosophical Transactions of the Royal Society of London.*

——— and Collingridge, G. L. 1993. A synaptic model of memory: long-term potentiation in the hippocampus. *Nature* 361:31–39.

Bonhoeffer, T. and Yuste, R. 2002. Spine motility. Phenomenology, Mechanisms, and Function. *Neuron* 12:1019–27.

Brodal, P. 1992. *The Central Nervous System. Structure and Function.* New York: Oxford University Press.

Carew, T. J. and Sahley, C. L. 1986. Invertebrate learning and memory: from behavior to molecules. *Annual Review of Neuroscience* 9:435–87.

Castellucci, V. F. and Kandel, E. R. 1974. A quantal analysis of the synaptic depression underlying habituation of the gill—withdrawal reflex in aplysia. *Proceedings of the National Academy of Sciences* 77:7492–96.

Changeux, J.-P. 2002. *L'homme de vérité.* Paris: Odile Jacob.

Cheung, V. G. and Spielman, R. S. 2002. The genetics of variation in gene expression. *Nature Genetics Supplement* 32: 522–25.

Childress, A. R., Mozley, P. D., McElgin, W., et al. 1999. Limbic activation during cue-induced cocaine craving. *American Journal of Psychiatry* 156:11–18.

Craig, A. D. 2002. How Do You Feel? Interoception: the sense of the physiological condition of the body. *Nature Reviews Neuroscience* 3.8:655–66.

Damasio, A. R. 1994. *Descartes' Error. Emotion, Reason, and the Human Brain.* New York: Putnam.

——— 2003. Feelings of emotion and the self. *Annual of the New York Academy of Sciences* 1001:253–61.

Drews, G. 1996. Endocrine pancreas. In *Comprehensive Human Physiology*, eds. R. Greger and G. Windhorst, 569–78, New York: Springer.

Edelman, G. M. 1992. *Bright Air, Brilliant Fire: On the Matter of the Mind.* New York: Basic Books.

Echtegoyen, R. H. and Miller, J.-A. 1996. *Silence Brisé. Entretien sur le mouvement psychanalytique.* Paris: Seuil.

Eichenbaum, H. B., Cahill, L. F., Gluck, M. A., et al. 1999. Learning and memory: systems analysis. In *Fundamental Neuroscience*, ed. M. J. Zigmond, F. E. Bloom, S. C. Landis, et al., 1455–86. San Diego: Academic Press.

Freud, S. 1887–1902. *The Origins of Psycho-Analysis. Letters to Wilhelm Fliess, Drafts and Notes: 1887–1902*, ed. M. Bonaparte, A. Freud, and E. Kris, trans. E. Mosbacher and J. Strachey. New York: Basic Books.

——— 1891. *On Aphasia. A Critical Study.* Trans. E. Stengel. New York: International Universities Press 1953.

——— 1895. Project for a scientific psychology. *Standard Edition* 1:281–397.

——— 1898. The psychical mechanism of forgetfulness. *Standard Edition* 3:287–97.

——— 1900. *The Interpretation of Dreams. Standard Edition* 4–5.

——— 1901. The forgetting of proper names. *Standard Edition* 6:30–32.

——— 1908. On the sexual theories of children. *Standard Edition* 9:205–26.

——— 1909. Analysis of a phobia in a five-year-old boy. *Standard Edition* 10:1–147.

——— 1910. *Leonardo da Vinci and a Memory of His Childhood. Standard Edition* 11:57–137.

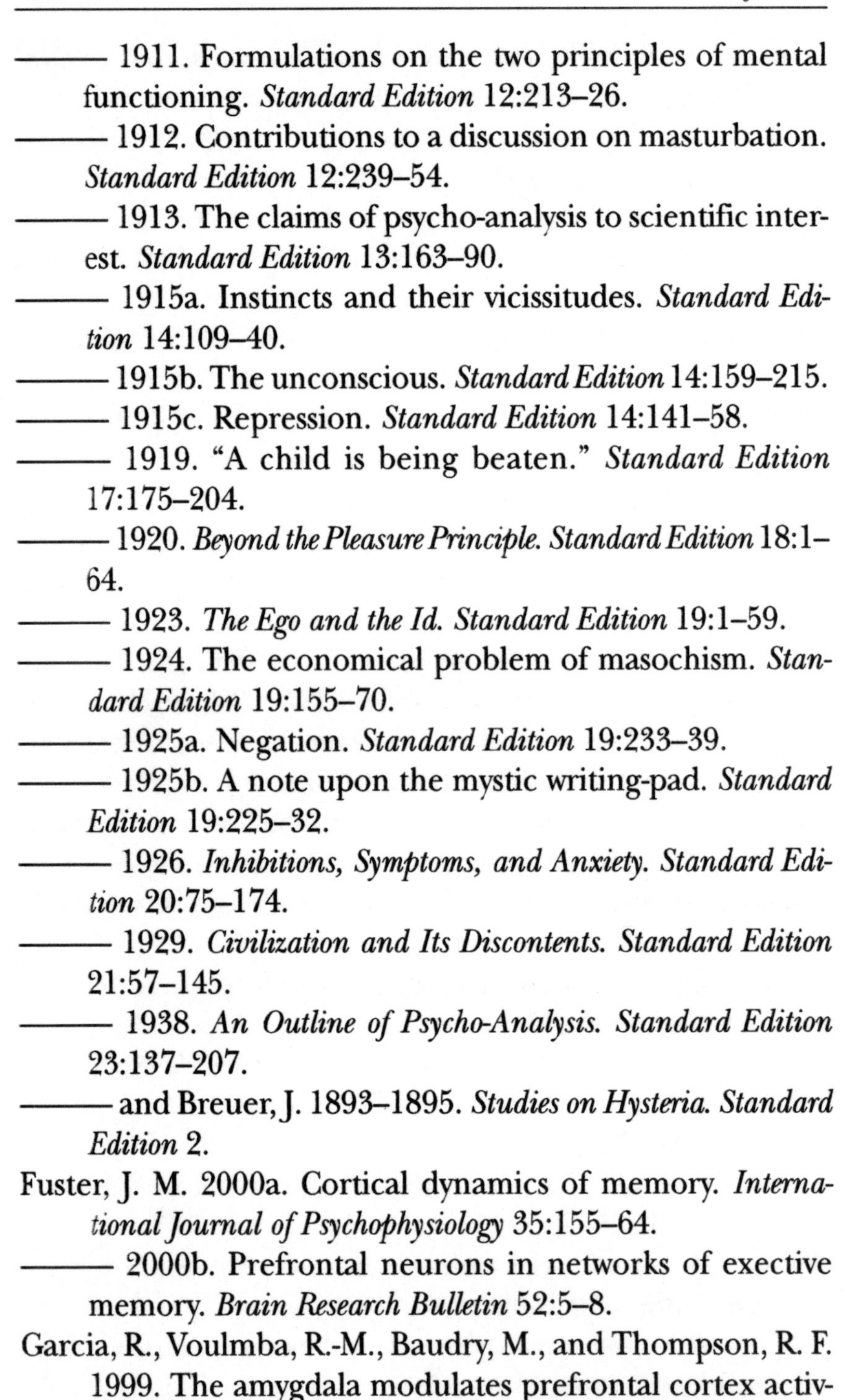

——— 1911. Formulations on the two principles of mental functioning. *Standard Edition* 12:213–26.
——— 1912. Contributions to a discussion on masturbation. *Standard Edition* 12:239–54.
——— 1913. The claims of psycho-analysis to scientific interest. *Standard Edition* 13:163–90.
——— 1915a. Instincts and their vicissitudes. *Standard Edition* 14:109–40.
——— 1915b. The unconscious. *Standard Edition* 14:159–215.
——— 1915c. Repression. *Standard Edition* 14:141–58.
——— 1919. "A child is being beaten." *Standard Edition* 17:175–204.
——— 1920. *Beyond the Pleasure Principle. Standard Edition* 18:1–64.
——— 1923. *The Ego and the Id. Standard Edition* 19:1–59.
——— 1924. The economical problem of masochism. *Standard Edition* 19:155–70.
——— 1925a. Negation. *Standard Edition* 19:233–39.
——— 1925b. A note upon the mystic writing-pad. *Standard Edition* 19:225–32.
——— 1926. *Inhibitions, Symptoms, and Anxiety. Standard Edition* 20:75–174.
——— 1929. *Civilization and Its Discontents. Standard Edition* 21:57–145.
——— 1938. *An Outline of Psycho-Analysis. Standard Edition* 23:137–207.
——— and Breuer, J. 1893–1895. *Studies on Hysteria. Standard Edition* 2.
Fuster, J. M. 2000a. Cortical dynamics of memory. *International Journal of Psychophysiology* 35:155–64.
——— 2000b. Prefrontal neurons in networks of exective memory. *Brain Research Bulletin* 52:5–8.
Garcia, R., Voulmba, R.-M., Baudry, M., and Thompson, R. F. 1999. The amygdala modulates prefrontal cortex activity relative to conditioned fear. *Nature* 402:294–96.

Gracián, B. 1651–1657. *The Critick.* Trans. P. Rycaut. London, 1681.

Goldman-Rakic, P. S. 1999. The physiological approach: functional architecture of working memory and disordered cognition in schizophrenia. *Biological Psychiatry* 46:650–61.

Greger, R. and Windhorst, U., eds. 1996. *Comprehensive Human Physiology.* New York: Springer.

Guttmacher, A. E. and Collins, F. S. 2003. Welcome to the genomic era. *New England Journal of Medicine* 349:996–98.

Hebb, D. O. 1949. *The Organization of Behavior.* New York: Wiley.

Héritier, F. 1996. Réflexions pour nourrir la réflexion. In *Séminaire de Françoise Héritier. De la Violence,* 13–53. Paris: Odile Jacob.

Humeau, Y., Shaban, H., Bissière, S., and Lüthi, A. 2003. Presynaptic induction of heterosynaptic associative plasticity in the mammalian brain. *Nature* 426:841–45.

Insel, T. R. and Collins, F. S. 2003. Psychiatry in the genomics era. *American Journal of Psychiatry* 160:616–20.

James, W. 1890. *The Principles of Psychology.* New York: Dover, 1950.

Jones, T. A. and Greenough, W. T. 2002. Behavioral experience-dependent plasticity of glial-neuronal interactions. In *The Tripartite Synapse Glia in Synaptic Transmission,* eds. A. Volterra, P. J. Magistretti, and P. G. Haydon, 248–65. New York: Oxford University Press.

Jungermann, K. and Barth, C. A. 1996. Energy metabolism and nutrition. In *Comprehensive Human Physiology,* eds. R. Greger and G. Windhorst, 1425–57. New York: Springer.

Kandel, E. R. 1998. A new intellectual framework for psychiatry. *American Journal of Psychiatry* 155:457–69.

——— 1999. Biology and the future of psychoanalysis. A new intellectual framework for psychiatry revisited. *American Journal of Psychiatry* 156:505–24.

——— 2001a. The molecular biology of memory storage. A dialogue between genes and synapses. *Science* 294:1030–38.

——— 2001b. Psychotherapy and the single synapse: the impact of psychiatric thought on neurobiological research. *Journal of Neuropsychiatry and Clinical Neurosciences* 13:290–300.

Kempermann, G., Wiscott, L., and Gage, F. H. 2004. Functional significance of adult neurogenesis. *Current Opinion in Neurobiology* 14:186–91.

Koizumi, K. 1996. The role of the hypothalamus in neuroendocrinology. In *Comprehensive Human Physiology*, eds. R. Greger and G. Windhorst, 379–402. New York: Springer.

Koob, G. F. 2003. Neuroadaptive mechanisms of addiction: studies on the extended amygdala. *European Neuropsychopharmacology* 13:442–52.

Krafft-Ebing, R. von. 1886. *Psychopathia Sexualis.* Trans. F. J. Rebman. New York: Kessinger 2006.

Kuhn, T. S. 1970. *The Structure of Scientific Revolutions.* Chicago: University of Chicago Press.

LaBar, K. S., Gatenby, J. C., Gore, J. C., LeDoux, J. E., and Phelps, E. A. 1998. Human amygdala activation during conditioned fear acquisition and extinction: a mixed trial fMRI study. *Neuron* 20:937–45.

Lacan, J. 1948. Aggressivity in psychoanalysis. In *Écrits. A Selection.* Trans. and ed. Alan Sheridan, 8–29. New York: Norton, 1977.

——— 1954. Réponse au commentaire de Jean Hippolyte sur la *Verneinung* de Freud. In *Écrits*, 369–80. Paris: Seuil, 1966.

——— 1954–1955. *The Seminar, Book II. The Ego in Freud's Theory and in the Technique of Psychoanalysis*, trans. Sylvan Tomaselli. Cambridge, UK: Cambridge University Press.

——— 1955–1956. On a question preliminary to any possible

treatment of psychosis. In *Écrits. A Selection.* Trans. and ed. Alan Sheridan, pp. 179–225. New York: Norton, 1977.
——— 1956–1957. *Le Séminaire. Livre IV. La Relation d'Objet.* Paris: Seuil.
——— 1959–1960. *The Seminar. Book VII. The Ethics of Psychoanalysis,* trans. Dennis Porter. London: Routledge, 1992.
——— 1969, 1964. Position de l'inconscient. In *Écrits,* 829–50. Paris: Seuil, 1966.
——— 1962–1963. *Le Séminaire. Livre X. L'Angoisse.* Paris: Seuil, 2004.
——— 1964. *The Four Fundamental Concepts of Psychoanalysis.* Trans. Alan Sheridan. London: Hogarth, 1977.
——— 1966. Allocution sur les psychoses de l'enfant. In *Autres Écrits,* 50–69. Paris: Seuil, 2001.
——— 1973. L'étourdit. *Scilicet* 4:5–52.
Lamprecht, R. and LeDoux, J. 2004. Structural plasticity and memory. *Nature Review of Neuroscience* 5:45–54.
LeDoux, J. 1994. Emotion, memory, and the brain. *Scientific American* 270:50–57.
——— 1996. *The Emotional Brain.* New York: Simon & Schuster.
——— 2003. Emotional brain, fear, and the amygdala. *Cellular and Molecular Neurobiology* 23:727–38.
Levy, B. J. and Anderson, M. C. 2002. Inhibitory processes and the control of memory retrieval. *Trends in Cognitive Sciences* 6:299–305.
Lüscher, C., Nicoll, R. A., Malenka, R. C., and Muller, D. 2000. Synaptic plasticity and dynamic modulation of the postsynaptic membrane. *Nature Neuroscience* 3:545–50.
Magistretti, P., Pellerin, L., Rothman, D. L., and Shulman, R. G. 1999. Energy on demand. *Science* 283:496–97.
Malabou, C. 1996. *L'avenir de Hegel. plasticité, temporalité, dialectique.* Paris: Vrin.
——— ed. 2000. *Plasticité.* Paris: L. Scheer.
——— 2004. *Que faire de notre cerveau?* Paris: Bayard.

Malenka, R. C. 2003. The long-term potential of LTP. *Neuroscience* 4:923–26.

Markram, H., Lubke, J., Frotsher, M., and Sakmann, B. 1997. Regulation of synaptic efficacy by coincidence of postsynaptic APs and EPSPs. *Science* 275:213–15.

Mattay, V. S., Goldberg, T. E., Fera, F., et al. 2003. Catechol O-methyltransferase val ^{158}-met genotype and individual variation in the brain response to amphetamine. *Publications of the National Academy of Sciences* 100:6186–91.

McDonald, A. J. 1998. Cortical pathways to the mammalian amygdala. *Progress in Neurobiology* 55:257–332.

McEwen, B. S. 1996. Hormones modulate environmental control of a changing brain. In *Comprehensive Human Physiology*, ed. R. Greger and G. Windhorst, 473–93, New York: Springer.

McGaugh, J. L. 2004. The amygdala modulates the consolidation of memories of emotionally arousing experiences. *Annual Review of Neuroscience* 27:1–28.

McNaughton, B. L. 2003. Long-term potentiation, cooperativity, and Hebb's cell assemblies: a personal history. *Philosophical Transactions of the Royal Society of London B* 358:629–34.

Miller, J.-A. 1987. Les réponses du réel. In *Aspects du malaise dans la civilisation. Psychanalyse au CNRS*, 9–22. Paris: Navarin.

——— 2000. Biologie lacanienne et événement du corps. *La cause freudienne* 44: 5–19.

Molière, J.-B. 1665. *Don Juan.* In *Don Juan and Other Plays.* Trans. G. Gravely and I. Maclean, 40–93. New York: Oxford University Press.

Morris, J. S., Öhman, A., and Dolan, R. J. 1998. Conscious and unconscious emotional learning in the human amygdala. *Nature* 393:467–70.

——— 1999. A subcortical pathway to the right amygdala

mediating unseen fear. *Proceedings of the National Academy of Science* 96:1680–85.

Morris, R. G. M., Moser, E. I., Riedel, G., et al. 2003. Elements of a neurobiological theory of the hippocampus: the role of activity-dependent synaptic plasticity in memory. *Philosophical Transactions of the Royal Society of London B* 358:773–86.

Nancy, J.-L. 2003. *Au fond des images.* Paris: Galilée.

O'Brien, C. P. 2003. Research advances in the understanding and treatment of addiction. *American Journal of Addiction* 12, Supplement 2:36–47.

Pavlov, I. P. 1927. *Conditioned Reflexes: An Investigation of the Physiological Activity of the Cerebral Cortex.* London: Oxford University Press.

Prigogine, I. 1996. *La fin des certitudes.* Paris: Odile Jacob.

Ramon y Cajal, S. 1909–1911. *Histologie du système nerveux de l'homme et des vertébrés.* Paris: A. Maloine.

Plato. *Symposium.* In *Great Dialogues of Plato,* trans. W. H. D. Rouse. New York: Mentor 1956.

Sandi, C., Merino, J. J., Cordero, M. I., et al. 2003. Modulation of hippocampal NCAM polysialylation and spatial memory consolidation by fear conditioning. *Biological Psychiatry* 54:6, 599–607.

Schmith, V. D., Campbell, D. A., Sehgal, H., et al. 2003. Pharmacogenetics and disease genetics of complex diseases. *Cellular and Molecular Life Sciences* 60:1636–46.

Shors, T. J., Miesegaes, G., Beylin, A., et al. 2001. Neurogenesis in the adult is involved in the formation of traces of memories. *Nature* 410:372–76.

Singer, W. 1998. Consciousness and the structure of neuronal representations. *Philosophical Transactions of the Royal Society of London B* 353:1829–40.

——— 2004. Synchrony, oscillations, and relational code. In *The Visual Neurosciences,* vol. 2, eds. L. M. Chapula and J. S. Werner, 1665–81. Cambridge, MA: MIT Press.

Smith, E. E. and Jonides, J. 1999. Storage and executive processes in the frontal lobes. *Science* 283:1657–61.

Solms, M. and Turnbull, O. 2003. *The Brain and the Inner World. An Introduction to the Neuroscience of Subjective Experience.* New York: Other Press.

Squire L. R., Stark, C. E. L., and Clark, R. E. 2004. The medial temporal lobe. *Annual Review of Neurosciences* 27:279–306.

Sylvester, D. 2001. *Giacometti.* Paris: Dimanche.

Van Praag, H., Christie, B. R., Sejnowski, T. J., and Gage, F. H. 1999. Running enhances neurogenesis, learning, and long-term potentiation in mice. *Proceedings of the National Academy of Science* 96:13427–13431.

Vernant, J.-P. 1999. *Myth and Society in Ancient Greece,* trans. Janet Loyd. New York: Zone.

Weinberger, N. M. 1998. Physiological memory in the primary auditory cortex: characteristics and mechanisms. *Neurobiology of Learning and Memory* 70:226–51.

——— 2004. Specific long-term memory traces in the primary auditory cortex. *Neuroscience* 5:279–90.